TQM-based Project Planning

TQM-based Project Planning

D. L. Kimbler and William G. Ferrell
Department of Industrial Engineering
Clemson University
Clemson
South Carolina, USA

CHAPMAN & HALL
London · Weinheim · New York · Tokyo · Melbourne · Madras

Published by Chapman & Hall, 2–6 Boundary Row, London SE1 8HN, UK

Chapman & Hall, 2–6 Boundary Row, London SE1 8HN, UK

Chapman & Hall GmbH, Pappelallee 3, 69469 Weinheim, Germany

Chapman & Hall USA, 115 Fifth Avenue, New York, NY 10003, USA

Chapman & Hall Japan, ITP-Japan, Kyowa Building, 3F, 2–2–1 Hirakawacho, Chiyoda-ku, Tokyo 102, Japan

Chapman & Hall Australia, 102 Dodds Street, South Melbourne, Victoria 3205, Australia

Chapman & Hall India, R. Seshadri, 32 Second Main Road, CIT East, Madras 600 035, India

First edition 1997

Typeset in 11/13 pt Palatino by Type Study, Scarborough
Printed in Great Britain at the University Press, Cambridge

ISBN 0 412 58860 9

A catalogue for this book is available from the British Library

Library of Congress Catalog Card Number: 96-85714

Printed on acid-free text paper, manufactured in accordance with ANSI/NISO Z39.48–1992 (Permanence of Paper)

To my wife, Beth

D.L.K.

To my mother, Lucille, and
to the memory of my father, Garland

W.G.F.

Contents

Preface

This book, bridging some of the tools associated with quality improvement and the process of project planning, is a solution to the synthesis of several problems. First, people tend to not plan well because that is easier in the short run. That is, good planning is sometimes not done because the planning process itself appears overwhelming. The planning process appears overwhelming because references and training in planning tend to concentrate on a few planning tools and their use, and implicitly assume that all the rest of the work will get done. People who need to plan projects and programs know this is not so, but don't know what to do about it. What is needed is more of a system view of planning, from the initial definition of the project through its review mechanism. We attempt to provide this system view.

A second issue is the apparently universal desire for a roadmap. In our teaching, research and practice in quality engineering, we repeatedly find enthusiastic people who would do good things if they could only figure out where to start and how to proceed. At one extreme we find the how-to-do-it works, the 'do these five things and all will be well' kinds of guides. Unfortunately, few worthwhile projects are simple enough to succeed through strict application of a template. At the other extreme, Dr W. Edwards Deming taught a philosophy and set of principles, in the conviction that the implementation must come from within the context of the organization. We present the execution of a project with a structure and set of examples that provide guidance but must be customized to the reader's context. It is not our intention to provide a strict step-by-step approach; even though we present a sequential structure, this is only one sequential structure, and it should be adapted as appropriate in each implementation.

Finally, the collection of tools from quality improvement used within the context of project planning addresses two issues. First, we encourage the use of these tools in activities not routinely associated with quality improvement. They are simple but powerful and informative, and their use should be more widespread than it is. Broadening their use is itself supportive of quality improvement, as it broadens the understanding of the systems we work within. Second, these tools come from a variety of origins and references. One of our goals is the presentation of these tools within a single context, where the understanding of their relationships may be clearer, and their use may be facilitated.

In a sense, this book is as much a book on quality improvement as it is on project planning. While the focus is certainly on large-scale projects, the principles and structure apply to process and system improvement projects and programs as well. Readers are encouraged to view project planning as an example context, and try to fit these structures, tools and models into other areas.

PART ONE

Introduction

Planning is one of the most important parts of any large-scale project of any significant complexity, yet it is typically the part that is done least well. In business a great deal is made of 'performance to plan', but frequently more effort is spent in measuring the performance than in developing the plan that causes it. A proverb in many cultures says that if we are completely lost, any direction is a good one. The equivalent in planning is to say that if our goals are unknown, any plan is a good one. But in any meaningful endeavor, our goals are not unknown. Having determined the overall goals of a project, a plan to accomplish them is absolutely necessary, and the success in attaining the goals will depend on the quality of the plan.

Some plans are strictly sequential. That is, the process to attain the goals consists of a set of steps which, executed in order, results in the successful completion of the project. While these plans may correspond to large projects, they are in fact simple in structure, even if they are comprehensive in detail. In more complex projects, things may take place simultaneously or even recursively. These make up for in complexity what they might lack in size. In both cases, however, the plan is important, and it is important to be able to go through the process of plan development.

The development of a plan for a project can seem to be an overwhelming task. Some things are unknown, others are loosely specified. It is tempting to just begin, and figure things out as we go. It is highly likely that this is the reason for poor planning, rather than the existence of doubts as to its necessity. If we try to develop a single plan, using a single method, to connect the start to the finish, this is an overwhelming task. If we consider, however, that planning is a set of activities, and that these activities can be done using a relatively simple set of tools, planning can become manageable. If we take a further step, and consider that many of these tools are already familiar, planning can be easy.

CHAPTER ONE

Planning and its place

So long as the conditions of the past remain in effect, we will continue to plan it that way.

(J.M. Juran, *Juran on Quality by Design*, 1992)

1.1 THE CONTEXT FOR PLANNING

The context for planning has changed, due in large part to the quality revolution that has been taking place around the world. The successes of the followers of Deming, Juran and others, in bringing about meaningful and substantial changes in product and service quality, also brought about changes in corporate culture and introduced new concepts and methods. The PDCA cycle (plan-do-check-act) originated by Shewhart (and frequently called the Deming cycle) is probably the oldest principle related to planning in the new quality science.

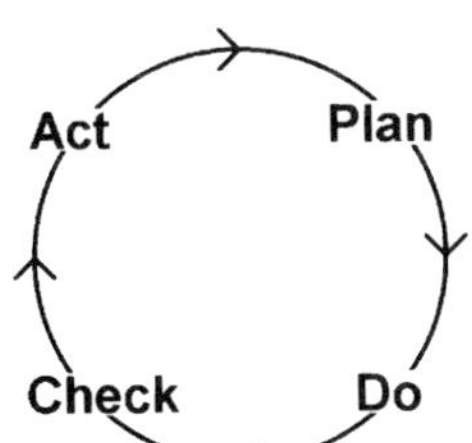

Figure 1.1 The Shewhart (or Deming) cycle

This principle, the basis of continuous improvement, explicitly requires making a change to a system, checking the results, taking action based on the effect and, most importantly, a plan.

Another version of continuous improvement comes from the Japanese word for improvement, *kaizen*. This incremental and continuous approach to system improvement, central to many Japanese industrial successes, has its own basis in planning. While it was initially greeted with skepticism by Westerners who tended to improve systems by breakthrough or major innovation, the recognition that innovation itself is amenable to planning has led to a combination of continuous improvement with breakthrough change, both the result of careful planning. The graph in Figure 1.2 shows the relative difference between incremental improvement, breakthrough improvement and their combination. Clearly, a combination of continuous incremental improvement coupled with planned innovation is a powerful thing.

The important point here is that neither continuous improvement nor breakthrough are reliable without appropriate planning. Breakthroughs happen, certainly, and sometimes they arise from serendipity. A focused effort, however, based on a plan with clear goals, can help us avoid relying on accidents.

1.2 VISION TO MILESTONES

The quality revolution has brought with it a new lexicon, and methods developed to implement the concepts of system quality

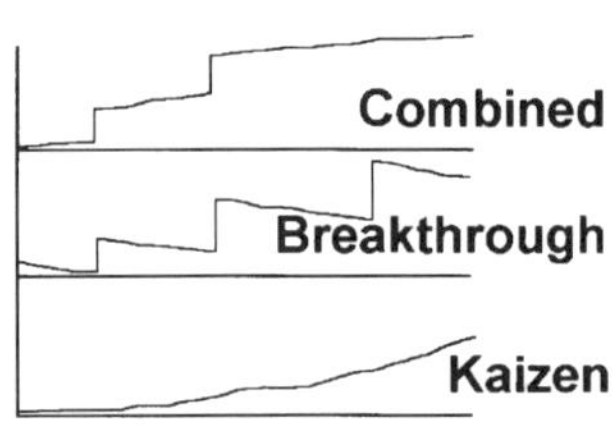

Figure 1.2 Breakthrough and continuous improvement

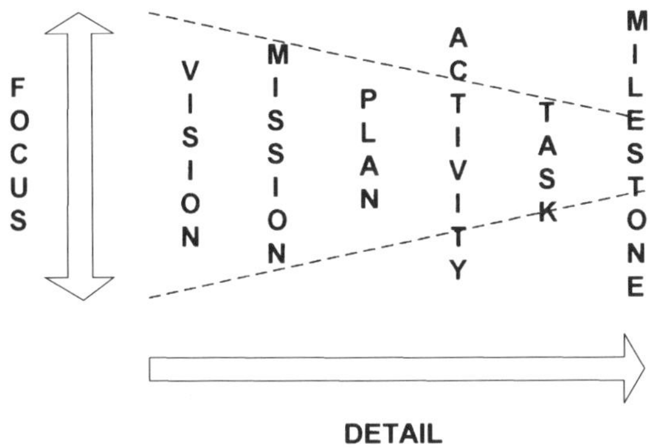

Figure 1.3 From vision to milestones

in the organizations affected. These concepts of system quality are much broader than the traditional methods of quality control. They deal with the organization in all its aspects, and cover organizational activities from conceptual through implementation.

Some of the key concepts are shown symbolically in Figure 1.3. As we move to the right in both increasing time and level of detail, we begin with a statement of vision. The vision statement is the corporate equivalent of the answer to the question 'What do you want to be when you grow up?' That is, the statement is broad, implies a future state, and tells what we (the organization) aspire to. The mission statement is also broad, but has a different flavor. Our mission is what we do, it is our immediate purpose. For example, the vision of a company that provides cleaning services might be to become the best in its class, in its service area. Its mission is to provide cleaning services of the highest quality to its customers.

This company has goals that are consistent with its vision and mission. Goals are still broadly stated, but are more action oriented. One of the cleaning service company's goals could be to expand its customer base substantially beyond its present market area. An objective corresponding to this goal is a quantitative statement, such as 'expand the customer base by 50% within the next two years'. The process of achieving this involves several activities, each identified as complete when its milestone is complete. In a large organization, this process of moving from vision to milestones is done at several levels, beginning with the top. Lower levels can then intelligently develop their vision, mission, etc., in the context of the broader organization.

It is clear that succeeding in achieving milestones results in meeting objectives, which causes goals to be satisfied. This is done within the context of the mission, and is aimed at realizing the vision. One distinction to make is that vision-oriented activities are strategic, while milestones are operational. Somewhere in the spectrum from vision through milestones, we switch from strategic thinking to operational thinking. Further, as we move downward in the organization we are also moving from strategic issues to operational issues. It is worthwhile to distinguish between strategic planning, dealing with long-term issues, and operational planning, dealing with short-term and action-oriented issues. Both of these become clearer within the context of quality management, which also provides some of the tools used in planning.

1.3 WHY PLAN?

Planning, as discussed in this book, is the process of defining actions to be taken to realize objectives. Some things are inherently easy to plan. Activities which are well-defined and concrete are typically easy, if the goals and objectives are clear. Planning a family trip is easy. Planning a military exercise or a corporate restructuring is not. One of the key differences is clarity of goals. Another is quality of information.

Consider the family trip. If we know where we are going and why, this is an easy task in the sense that we know how to do it. (It might still be complicated in terms of details.) If we don't know where we are going, planning is difficult. While this might seem obvious, consider how many organizations restructure themselves without a clear statement of goals, or with inappropriate goals, or without the understanding of the people involved. When the restructuring does not yield the desired results, managers are surprised, but should they be? If they began without a clear plan, they should be more surprised with success.

When good things happened in the absence of a good plan, they were accidents! They happened in spite of our design, rather than because of it, and they will happen again (or continue to happen) only by accident. This, more than any other, is the reason for good planning. It allows us to direct actions to achieve desirable outcomes, rather than tempting fate.

1.4 INTRODUCTION TO THE EXAMPLES

The examples used to illustrate the key concepts in this book trace the trials and tribulations that one division of the huge conglomerate F.A. Adams, Inc. (Big Board symbol: FAA) face as they try to recapture lost market share. Specifically, the Paper Airplane Division (PAD) is the largest producer of paper airplanes in the world. PAD sells high precision planes to retailers throughout the US and enjoys an excellent reputation with them. As recently as five years ago, PAD so thoroughly dominated that they effectively had 100% of the market and there were no competitors on the horizon; however, things quickly changed. A foreign specialty toy manufacturer and a domestic paper company formed a joint venture company to compete directly with PAD and the capitalization was sufficient to allow new machinery to be purchased and talented engineers to be hired. The result has been dramatic. This competitor has been steadily gaining market share every year and now services 30% of the market. A new general manager was recently appointed to 'straighten this mess out and reclaim our territory' and immediately drafted a trusted friend, R. Pisces Lott (her friends call her Pi), to take care of the technical side of the problem.

Pi possesses a basic understanding of TQM, and immediately identifies her first task as a road trip – that is, go and visit a few former customers at their locations to gather information. She fully appreciates that the only way to completely define the technical task that lies ahead is to understand the customer's concerns and communicate this insight to the employees at PAD. At the conclusion of two full weeks on the road, the conclusions were not unexpected. Her visits had been quite productive, although not very comforting. All of the customers expressed the same disturbing news, the quality of the competitor's product was simply better. Each had personal experiences to support the claims, however, the anecdotal evidence was very inconsistent from customer to customer and embodied a wide range of problems. Pi initially thought that maybe these customers were overly critical and just looking for any excuse to justify their actions.

Upon careful reflection, she decided that although she couldn't immediately put her finger on the technical problem, it was very unlikely that this many formally loyal customers would be creating stories and a much better explanation was that there were real problems that needed to be viewed as opportunities. Specifically, she couldn't help but think of PAD's *modus operandi* of fighting fires whenever and wherever they flared, and the big laugh one of her colleagues got when he bought a real fireman's hat and put it on his filing cabinet for everyone to see. 'This time', she thought, 'we're going to avoid the temptation of quick fixes and attack the root causes.'

It so happened that her search for answers led her to call a former faculty member who sent her the latest draft of his book on TQM-based Project Planning. Pi begins to read the book and can't stop. About 4 a.m. she finishes and realizes that these are exactly the techniques she needs to create the plan for improving quality and recapturing market share. She is so excited by the idea that she doesn't even attempt to sleep, only showers and heads for work. By the time the business day starts, she has a tentative plan on paper and gains the general manager's approval before the first cup of coffee.

The process

Manufacturing high-volume, high-precision, relatively low-cost paper airplanes is not a trivial matter. The problem is further complicated by the fact that, in some cases, the same plane must be produced to different specifications for different retailers.

The basic process of forming a paper airplane is the same for all models. The raw material consist of rolls of paper from which the appropriate pattern is cut. Any decorative printing or decals are applied to the flat pattern and, then, a series of folding machines creates the basic shape of the plane. The next process consists of applying a high-strength, quick-setting epoxy at critical points to maintain the shape. Finally, the entire plane is sprayed with a polyurathene sealer and 'baked' until it hardens. Final touches (e.g. weights, turning up the wing tips, etc.) are then added and the planes are packaged and shipped. The shipping carton resembles a large egg crate with each plane being stored in its own compartment.

The service aspect

Frequently the manufacturing sector and service sector are discussed as if they were two distinct entities. In fact, they share many characteristics and, even when their product is different, some common tools can be brought to bear on each. Manufacturing itself has many service functions within it. As the technical group proceeds to plan its recapture of market share, the general manager is also beginning to focus attention on the support that manufacturing and other groups will need. The computer support area has been selected as a critical element in the support of various company areas that will be involved in the broader project.

1.5 SUMMARY OF KEY POINTS

- PDCA – Plan, Do, Check, Act.
- Define the planning context with clear statements of vision and mission.
- Distinguish between strategic planning and operational planning, and don't try to do both the same way.
- Don't have your future determined by accidents.

CHAPTER TWO

Learning planning

> [The problem of productivity] will only be remedied when we learn how to manage people in such a way that they can work together more effectively.
>
> (William Ouchi, *Theory Z*, 1982)

2.1 GETTING STARTED

Good planning requires a good environment and the right tools. The previous chapter described the necessary understanding of vision and mission. One of the most important factors in the environment is constancy of purpose, appropriately emphasized by Dr W. Edwards Deming as the first of his 14 points. Constancy of purpose in the management of the organization provides the stability necessary for good planning to take place. It is difficult enough to begin a complicated project which itself may contain unknown factors; we should not add to our difficulties by imposing them internally.

The right tools help decompose the large and complex task of planning into a set of logical, do-able activities. In fact, one of the greatest hindrances to good planning is ignorance of simple and useful tools. Planning is a process, and it is amenable to organization, using some of the same tools used in continuous quality improvement. With this approach to planning, the planning process integrates well with other continuous improvement activities, because they share the same information and methods.

Table 2.1 The Seven Quality Control Tools

Checklist
Histogram
Pareto chart
Scatter diagram
Run chart
Cause and effect diagram
Statistical process control

Table 2.2 The Seven Management and Planning Tools

Affinity diagram
Tree diagram
Influence diagram
Matrix diagram
Activity network
Process Decision Program Chart
Matrix data analysis

The classical quality improvement tools can easily be regarded as planning tools. The Seven Management and Planning Tools (sometimes called the Seven New Tools) are appropriate for the initial stages of a project. The Seven Quality Control Tools (Seven Old Tools) are brought to bear as the project detail begins to take form. Tools from these common tool sets, except for those associated with quantitative process measurement and design (matrix data analysis and statistical process control) can be easily and productively applied in project management.

Other tools that have been found useful in quality improvement are also brought to bear. The concepts of policy deployment are vital to good execution of the project, and they form part of the structure of planning. The deployment flowchart is found to be a versatile and useful tool, especially in documenting processes involving several components of the organization. We should keep in mind that the objective is good planning followed by good implementation; any tool that works should be considered. The tools in this book are presented as recommendations, within a recommended framework and sequence; specific circumstances will call for specialized tools.

2.2 PLANNING TO PLAN

The planning process shares other aspects of continuous improvement, beyond the use of common tools. The approach to planning can be made completely consistent with common approaches to continuous improvement. The use of multifunctional teams, for example, is as necessary in planning as it is in quality improvement, especially when the planning task spans departments or functions. Planning is not something to be left to experts in planning; the relevant experts are found in the processes affected. They provide the personnel for the planning teams, to complement the strategy and vision coming from corporate management. In this respect, planning teams are no different from continuous improvement teams.

Since the planning teams will be cross-sections of the organization, we should not expect all members to be equally expert in the methods used. Here we apply the same principle of just-in-time training that is so effective in other endeavors. Teams begin working, and are provided the necessary training, in the context of the project at hand, at the time the training is needed. This can be guided by good facilitation, which will also be needed to help the teams maintain focus, especially in the initial stages.

The management of detail will vary through the project, but it must be considered at the outset. Various records will need to be kept of team activities and their results. The level of detail will vary, but this record must be maintained as both a record of the planning activities and as its product, providing documents to guide the execution of the project.

2.3 SUMMARY OF KEY POINTS

- Develop and maintain an appropriate environment with constancy of purpose.
- Provide training in planning tools as needed.
- Provide good facilitation to maintain focus and guide teams.
- Manage team details that will become the plan.

PART TWO

Project Definition

INTRODUCTION

The motivation behind development of TQM-based Project Planning is a desire to assist project managers in avoiding an all too common experience, namely, hearing that little internal voice at the completion of a project saying, 'This project *could* have achieved so much more and really provided tremendous value for the company if only we had known about those few key items and been able to do some things differently.' Frequently, this is called '20–20 hindsight', with the implication that only after the project is complete can these issues be understood. If we define success as achieving the greatest number of objectives with minimum cost, the basic idea that a project must be completed before it is possible to see the best path that should have been traveled is fallacious.

The key is in dedicated and focused project planning and one technique to accomplish this is TQM-based Project Planning as outlined in this book. These techniques not only focus on identifying a wide range of objectives but also on organizing them in a way that a project manager can choose the paths that impact the largest number of objectives and provide the highest likelihood of success. The initial phase of project planning, then, centers on defining the objectives of the project and molding these objectives into a form that is amenable to project monitoring and control. In many projects, this aspect has been trivialized, especially if the technical goal is well defined (e.g.

build a bridge) and all other objectives can be loosely aggregated in the charge, 'Get it done on time and under budget!' In this environment, the project manager actually spends little time performing 'project management', rather, he/she is consumed with a barrage of changes (e.g. engineering change orders, 'ECOs', in the contract engineering world) or pestering the team of technical experts to meet the deadlines and stay under budget. Decisions are frequently, if not exclusively, made in response to one crisis after another. Unfortunately, this virtually ensures that there is little coordination between resolutions, hence, the inconsistencies frequently breed more crises.

There are few project managers who can't relate to this. It is particularly frustrating to retrospectively look at a completed project and think about how much more could have been, and should have been, accomplished with the resources and energy expended. Another way to view this is that there is tremendous potential in most projects that never gets realized. The potential is nearly always present to accomplish more than the minimums but the key is gaining the additional insight and information at a point in the project when the critical decisions are made, not after the project is completed. TQM-based Project Planning is designed to tap this potential through a directed, thoughtful, and comprehensive planning process. It takes more time and energy at the start than traditional methods; however, this extra time and energy pays extraordinary dividends after the project is completed.

An important component associated with a successful project using TQM-based Project Planning is for the project team, including technical and managerial personnel, to be actively involved from the outset. It is equally important that the team be composed of members with diverse backgrounds because a variety of experiences and education forms a rich resource to extract information to be used in the planning effort. TQM-based Project Planning facilitates an organized approach to analysis and prioritization of this diverse information. The fundamental tone of the approach is dramatically different from historical project planning in which a non-trivial amount of time is spent in meetings devoted to arguing about 'on time and under budget', while important technical enhancements are never addressed.

Certainly when a diverse project team is asked to provide input into defining objectives, a large number of ideas will be generated and the content will span a wide spectrum of possibilities, even to the point of being considered wild, crazy, ridiculous or totally absurd. These ideas are not only acceptable, but encouraged at this point. Now is the time in the planning process that all ideas for project objectives are equally valuable. The purpose of this phase is to create a structure so that these merits of these ideas can be thoughtfully considered based on advantages and limitations, not knee-jerk opinion or comparison with the way it has always been done. Hence, this initial phase focuses and organizes the objectives and ideas into a form which conveys the proper emphasis and which is amenable to analysis, aggregation and prioritization.

In summary, TQM-based Project Planning is founded on the concept that the system requirements drive the planning effort rather than constraints associated with a particular planning model or mind-set forcing imposition of system requirements. The approach might look new and different, even a bit threatening; however, a bit of practice will breed confidence and understanding of the intrinsic power of this approach.

CHAPTER THREE

Definition of objectives

Once a company knows what is needed, it must focus people and jobs throughout the company according to what they can contribute to those needs.

(Shoji Shiba *et al.*, *A New American TQM*, 1993)

3.1 PLANNING OBJECTIVE: DEFINING THE KEY FACTORS

One of the fundamental differences in TQM-based Project Planning and more traditional approaches is that the objectives of the project are an integrated mix of requirements dictated by external agents and internal goals developed through this planning process. Traditional project management is typically geared solely towards meeting the external requirements. As might be expected for a technique grounded in total quality techniques, the emphasis here is to not only meet these specifications but achieve far beyond these expectations. A somewhat crude analogy can be drawn between comparing traditional PM methods and TQM-based Project Planning with a manufacturer who satisfies all customer requirements and one who actively seeks to delight the customer. A central premise is that minimal compliance is not sufficient to ensure success in today's business climate.

To achieve these ends, the first critical element is assembling the team that will create the plans to maximize the output of the project while ensuring that the minimum, external project requirements are met. Members of the project team must, necessarily, have diverse backgrounds and be knowledgeable, comfortable and confident with their discipline. This is absolutely necessary because these differing perspectives are the driving force behind the creation and synthesis of project objective(s) into both primary goals and subordinate goals. It is this diversity that will be tapped to create a plan to meet all the primary goals and as many subordinate goals as possible. One important element associated with successful completion of this task is for the team members to communicate using a common set of definitions and terms. The planning objective of this phase is to develop this base from which all can communicate. The base will consist of the objective(s) as well as key factors and ideas surrounding these objective(s). In real projects, however, the total number of ideas could easily become so large that further manipulations are impossible from a practical viewpoint.

To address this problem, the second objective of this phase seeks to locate patterns in these ideas and objectives that can be coalesced into a manageable number of major topics, typically not more than ten. The effort required to distinguish these patterns is not only done to reduce the number of key factors, but, more importantly, to allow individual members to recognize how the language of their specialty meshes with others and where their input is unique.

Taken together, these two planning objectives produce a base from which the team will work throughout the TQM-based Project Planning effort and help the team members understand some of the elements they have in common, and which are different, with others.

3.2 TOOL: AFFINITY DIAGRAM (OR KJ METHOD)

3.2.1 Description

The affinity diagram is a TQM tool used to organize bits of data/information that appear to be in chaos. The basis for aggregation is 'natural affinity', a concept that is rooted in the intuition and experience of the team. That is, the affinity diagram is constructed by having the team group sets of ideas that appear to have some characteristic in common. Affinity diagrams are commonly used to force a team to look at issues from different perspectives and generate creative solutions to old problems. In this application, project objectives and goals are first generated by decomposing external requirements and conducting brainstorming sessions into very small, elemental units of information. These units are subsequently aggregated using an affinity diagram that does not require a complete resolution to the problem, simply that similar ideas be aggregated. The output of this phase will likely begin to combine both novel and different ideas for the project with well-known, traditional objectives. The concept is that complex objectives must be decomposed, if possible, into elemental components and new ideas be single-issue as opposed to multi-faceted so that constructing the affinity diagram will be more efficient. Since the affinity diagram is well suited for aggregating and combining ideas, a large number of entities can easily be accommodated and, in fact, is encouraged. It is easier to combine a large number of single ideas than decide the fate of a few which have multiple aspects.

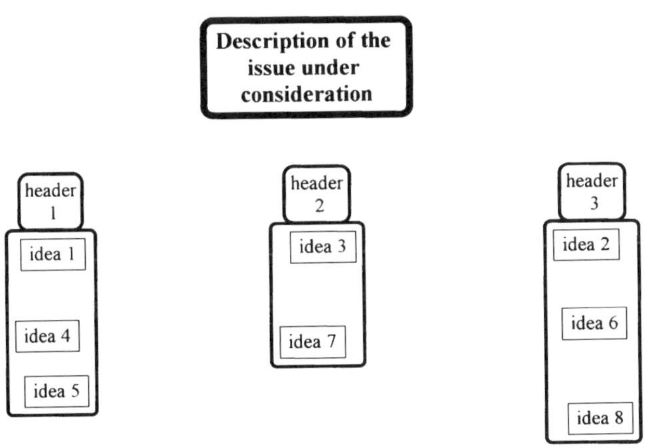

Figure 3.1 Example affinity diagram

There are a number of ways in which the objectives can be specified. It is recommended that a number of these be used because they will each generate different kinds of ideas. A technique to accomplish this is to task each team member with generating ideas individually or through one-on-one discussion. For example, interviewing a variety of people who have expertise in some aspect related to the project but who are not involved directly on the team is an excellent way to generate ideas. Members should be encouraged to speak with people who have tremendous expertise in what appears to be a tangential (or even unrelated) area and see if they have ideas that can be of value. At this point in the planning process, it is not acceptable to even think about eliminating ideas because 'we haven't ever done it like that before'. Another important element in the individual portion of the exercise is personal contemplation and reflection. With the current emphasis on 'teams', the value of having individuals ponder the issue at hand and introspectively consider alternatives is frequently lost, and it should not be. Individuals can often achieve a tremendous amount of creativity and synthesis when left alone with their thoughts.

After the individuals have been permitted to work independently, brainstorming is frequently used to integrate the results of the individual oriented exercises and the synergy of the group is used to expand and develop ideas. Although industry is quite familiar with brainstorming, it is worth remembering that there are three tenets associated with proper implementation: (1) no criticism is allowed so everyone is free to express any idea, (2) positive additions and modifications are allowed only with consensus and (3) the more ideas, the better. When a good project team is assembled and their members given proper time for individual contemplation and discussions with associates, an enormous creative potential has been intrinsically built into the system. Proper use of brainstorming should be seriously considered for harnessing these ideas and capturing them for input into the affinity diagram. The output of this phase is a large, and potentially huge, list of ideas that are single-issue in scope. Note that it is essential to decompose multi-issue ideas or the aggregation process will become quite difficult if not impossible.

The affinity diagram is used here to aggregate these single issue elements according to 'natural' affinity. 'Natural' is much like 'beauty', it can only be defined by the beholder and, in this case, the beholder is the team. The aggregation takes place without discussion or explanation and the output will be both clusters consisting of a number of ideas with a common theme and a few single ideas which do not appear to relate to anything else. The team is now instructed to document the progress.

Specifically, they must first discuss the details of each cluster including the general theme and why specific ideas are included and others omitted. Then, refinement is permitted where necessary and the details of the final affinity diagram are documented in a written report.

3.2.2 Construction

While many authors have defined the steps in constructing an affinity diagram, this section addresses the key steps in one implementation that works well in the project planning environment.

1. Before the brainstorming session, gather and develop some ideas and thoughts concerning the project and the stated objectives. A very successful technique is to charge the team with individually developing ideas through both interviews and private meditation. The ideas generated here can be further developed and validated using a brainstorming session. The ultimate purpose of this step is to develop a variety of ideas concerning the objectives of the project.
2. Record the concepts on individual cards or 'Post-It' notes. (The 'Post-It' notes are really handy for this exercise, although regular index cards and tape will work.) It is best to record one single-issue element on each note rather than a multi-faceted concept. Keeping the concepts simple aids the grouping that is performed later.
3. Arrange the cards so that ideas which are 'similar' are

grouped together. The key here is that all team members participate and the exercise must be done silently. There will be plenty of time to discuss the logic of the groupings later but, at this point, there should be no discussion. Some care should be taken so that the cards are initially distributed in a random sequence. If Post-It notes are used, a bare wall provides an excellent work space.

4. Label the groups of cards either by dividing the team into small groups and assigning each group of people an aggregated cluster of ideas or have the entire team label each cluster. If there are cards which do not seem to fit, they can be removed and remain by themselves or reassigned but do not force them into groups. 'Lone wolves' are quite fine. The label must be descriptive but short and convey exactly what the cards in that group describe, no more and no less. Abstractness is not a virtue in this phase and the labels are placed at the top of the stack. Completion of the labeling exercise indicates that the ideas have been grouped into at most ten groups according to some logic and appropriate labels have been created. Further, any individual cards have been carefully considered and are thought to be true individuals.
5. Clusters of ideas and the individuals that have some relationships are then aggregated by forming a cluster of the groups. The groups formed in step 4 are not disassembled – each packet of ideas is simply assigned to a larger group, where logical. This step may also be done in silent, although it is not as critical here as it is in the initial groupings. The completion of this step results in: clusters of the ideas grouped together, clusters grouped with individuals, individuals grouped, and both clusters and individuals that do not lend themselves to further grouping. We call this output, the final grouping.
6. Each member of the final grouping is then discussed. This is the time when the logic used to form the member is discussed along with each of the individual ideas. This, too, is a time where any additional thoughts should be vocalized and discussed. Logistically, small groups of team members should be assigned as the owner of one member of the final

grouping. Ownership requires taking notes on the discussion of that member and writing the final report, step 7. The output of this phase should be a written diagram of the final grouping and the individual ideas contained therein.

7. A written report on each member of the final grouping is constructed. The collection of the diagram along with these reports form the final affinity diagram. The report can contain objective statements ('It is . . .') along with subjective and speculative notions ('I think . . .' or 'It appears . . .').

Construction of the affinity diagram is required because this TQM-based methodology is based on the belief that the plan which will generate the best possible project must gather much more input from many more people that any traditional approach to project planning.

The affinity diagram transforms this wide variety of chaotic information about the objectives of the project into not more than ten groups of ideas, each bonded by some intrinsic relationship. It is important to recognize that this phase does not resolve all questions nor lay out a clear and unambiguous path to project planning, but remember this is only the first step.

3.3 SUCCESS FACTOR – CHOOSE A GOOD THEME

Choosing the project theme or task description is a very important key to ultimate success. Consider that the theme is the pivot around which the entire planning and work effort will revolve. Further, many substantial advances are discovered by those who think, or are allowed to think, outside the regular structure of a particular problem. They simply look at things differently and better resolutions become obvious. Clearly, the initial use of an affinity diagram supports collecting divergent opinion which is, as mentioned previously, one of the tenets of the TQM-based approach. Theme selection can be the catalyst to advances or the killer of all creative input. The following story accompanied early efforts at brainstorming and provides an excellent example about theme selection.

A local, all-male civic organization was planning fund-raising events for the year. Most of the members were beginning to outline what they had always done when one slightly different fellow said, 'It's too bad we aren't women because prostitution pays a lot better on an hourly basis.' After a round of laughter, one of his 'status quo' colleagues said, 'He's got a point, but I'm afraid my wife might not cozy to the idea in the traditional sense. On the other hand, my father taught me how to do simple plumbing while I was growing up and I still change washers and keep the drains clean around my house. I wonder if someone would pay for me to do that around their house and not have to pay a plumber for a house call?' It didn't take long for this radical idea to be developed into a year-long fund raising effort which netted incredible amounts of money and helped many an elderly person with common house problems at a fraction of the cost they would normally have had to pay. The key, however, was that the gentleman with the original idea thought the theme was 'how do we make money', and not 'on what date this year are we having our same old fund-raisers'. The theme makes all the difference. It can spur creative and diverse suggestions or restrict discussion to what has always been done.

3.4 MANUFACTURING EXAMPLE

The first task in developing the 'objectives' is to select the project team. Historically, a project team dedicated to the task of improving quality would have consisted solely of manufacturing personnel; however, the concepts in this chapter indicate that obtaining a wide spectrum of ideas is critical to maximizing the effectiveness of the process. Pi chooses a team consisting of the manufacturing manager, a process engineer, an airplane design engineer, a marketing representative, one outside sales person, and five people from operations (one foreman, one lead operator, one operator and two mechanics). Clearly, these people bring dramatically different viewpoints to the problem; however, a bit of reflection indicates the value of having each on the team.

Two weeks prior to the first meeting, Pi sent each team member an information package that included the report she wrote after visiting with the customers, not as a means of restricting their input but, rather, to provide insight into the customer's perception of the problems. Further, she clearly indicated that the overall objective of the project was to take a comprehensive look at improving the quality of the paper airplanes and the initial meeting would be devoted to brainstorming ideas concerning the task at hand. After thinking about the problem for the two weeks, the brainstorming session yielded the following list of ideas:

- better printing;
- make the planes look more like real planes
- cut the price so the customers can buy new ones more often;
- make the planes more durable and raise the price so the customers don't have to buy new ones very often;
- try radically different designs;
- reduce the cost of manufacturing;
- market what the customer wants, not what we want to make;
- reduce the cost by firing managers;
- achieve better dimensional stability;
- reduce defective planes that customers receive/inspect the final product;
- improve printing resolution;
- improve paper quality;
- prevent the print from rubbing off in the customer's hands (this happens under certain conditions but no one knows what those conditions are);
- change the glue to make them stick together longer;
- improve the process by which the over spray is applied;
- offer a wide variety of colors.

After all ideas had been exhausted, the team was tasked with constructing the affinity diagram in search of the 'objectives'. Figure 3.2 illustrates the final affinity diagram they constructed.

At this point, the focus of the example is narrowed to one 'objective', namely, the printing operation. The sole purpose of picking only one topic is to reduce the amount of repetition associated with the example through future chapters.

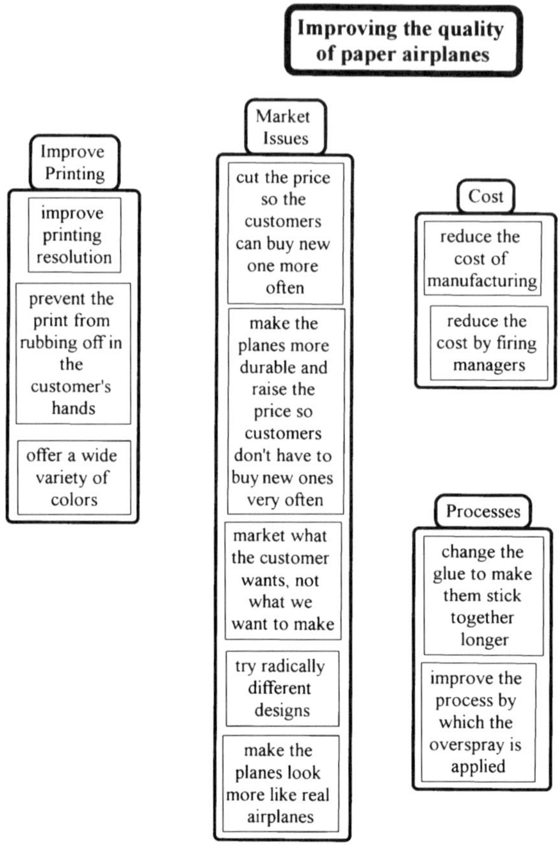

Figure 3.2 Affinity diagram on quality of paper airplanes

As prescribed in this chapter, the team took the final affinity diagram and provided written documentation which both detailed and sought to explain the results. The portion of the report dealing with the printing objective is contained below.

Improve printing
It appears that printing is a vital area for both quality improvement and for making advances in gaining new customers because the breadth of the ideas was tremendous. We think there are three major areas in this objective:

1. Expand the range of colors
Suggestions ranged from making them more faithful to real aircraft to making 'neon' colored planes.

2. Improve the resolution
Several ideas suggested that the print resolution was not comparable with our competitor and should be improved. We think it's important to note that the raw paper must be internally purchased from our paper mill. That is, we can't buy other people's paper nor do we appear to have much control over our internal supplier.

3. Make the ink permanent
Some customers report that the ink rubs off onto hands and clothing during 'normal handling and use'. The fact that this happens has turned into a huge advantage for the competitor and simply must be stopped permanently.

The team takes a final look at the 'objectives': improve printing (e.g. to improve the perceived way the planes look), improve manufacturing quality (e.g. quality), explore different designs (e.g. variety), creative marketing strategies, and cost control. They decide that these 'objectives' represent an ideal set for accomplishing the ultimate goal of improved quality, customer satisfaction, and increased market share.

3.5 SERVICE EXAMPLE

Just as PAD is beginning to take a continuous improvement approach to its manufacturing processes, it finds similar needs in the components of the company that provide service. Process improvement in providing service is not limited to the service sector; all organizations find themselves in the service business, if only to provide services internally.

The service sector in PAD of most immediate concern is that of providing computer support. Although not all companies have integrated computer support and networking into their processes to the extent of some of the industry giants (such as

Boeing), a great many depend on computing services to some extent. PAD is typical of companies of its size, in that its internal computing service is evolving. While some companies are evolving from a large centralized mainframe oriented service to a distributed service, others are evolving from small numbers of individual desktop computers into a distributed network.

PAD is in the second group. Not many years ago virtually everything but accounting was done manually. Design, production planning, manufacturing, all were done with paper, calculators (and before them, slide rules), and physical files. As is typical, certain functions began to be done using desktop computers, usually due to the interest and expertise of individuals. Over the years these functions have expanded with little organization or planning.

Today, PAD finds itself with a central network that was designed primarily for administrative support and a collection of computers that are individually configured for their non-administrative functions. The new general manager is concerned that the lack of integration of computing resources could stand in the way of his other initiative in improving the firm's competitive position. He is also very aware of the growing interest in the use of electronic mail and documents within PAD, as well as the potential use of external networking in marketing. The time appears to be ripe to expand this internal service.

The general manager's approach here is similar to the manufacturing problem. He has drafted another friend, Norman R. Nette (NR to his friends), to manage this transition. NR has been charged with determining an appropriate level of computing support for key functions within the company, and then coordinating the development of internal systems to provide this support.

NR's first step is to put together a small team representing the major areas of the business. This team identifies as its first order of business an assessment of the current status of computing support. The team members each review their areas of responsibility, to ensure that they have sufficient knowledge of the current status. While they may take this opportunity to solicit other information from their areas (such as need for new

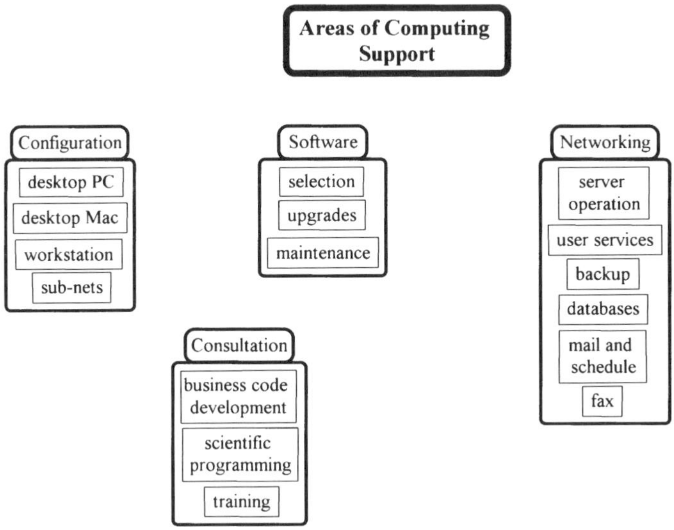

Figure 3.3 Affinity diagram on areas of computing support

services, or areas of inadequate support), the initial focus is on the baseline level of support.

The first meeting of this group served to communicate the charge to the team, establish housekeeping details and principles of operation, allow for discussion of the overall project, and schedule the next meeting with assignments for reporting back. The second meeting follows the individual data collection, and the team has adopted an affinity process to merge their findings. The affinity diagram in Figure 3.3 is developed in this meeting.

This affinity diagram shows the major functions presently performed by the computing support group, although not in the form of the present organization. The Configuration function is a service that is provided upon request to individuals and offices that need help in developing systems to meet their needs, and it ranges from individual workstations to small office networks. The Software function has been very informal, and has consisted again of providing software advice on request. The Networking function has been in existence for some time, and has primarily supported the administrative area. The Consultation function is another informal one, provided mainly to the more advanced users. Training has been found almost non-existent.

The affinity diagram is best regarded as a starting point. The fact that it does not duplicate the present organization is not a matter of great concern at this point; it is more to the point that this may highlight opportunities in modifying the organization. The varying levels of support and experience in the various areas are also not prescriptive. The object here is to develop a description of the baseline system, without being bound by its present structure. The affinity process lends itself well to this, providing a new view of computing support based on what the team members have observed. What is most significant about the affinity diagram is that it gives the team a rational framework for further work.

3.6 SUMMARY OF KEY POINTS

- Carefully define and state the issue under consideration.
- Allow ideas to flow freely.
- Look for new ideas and themes within affinity groupings.
- Don't try to solve the problem or generate the plan; we are just beginning to gather and organize information.

CHAPTER FOUR

Organizing objectives

One of these days we have to get organized.'
(Ubiquitous semi-humorous cartoon caption)

4.1 PLANNING OBJECTIVE: ORGANIZING THE OBJECTIVES

In the previous chapter, key objectives were developed by first defining elemental, single-issue components from diverse sources and, then, aggregating these components into natural groupings. Thus, the aggregated groupings reflect some intrinsic characteristics about the elements that bind them together. The process that created these groups renders it highly unlikely that there exist any additional, obvious themes between the elements (or the aggregates). Further attempts to find themes will likely be fruitless or, worse, groupings will be 'forced' along lines that have no logical basis. For the remainder of this chapter, we will refer to the groups of elements along with the single elements that could not be grouped as objectives.

It is interesting to note that some of the titles associated with the objectives will frequently sound very traditional, like 'make the widgets blue' or 'improve quality'. Others may reflect statements of business reality such as the fact that 'internal suppliers must be used' or that 'total production yield must exceed 40% to be profitable'. These should be considered normal and considered with equal importance as the others. On the other hand, if all the groupings reflect commonly understood and appreciated issues, it would be wise to review the idea-generation phase to ensure that sufficient diversity was considered.

TQM-based Project Planning now begins to transform the data into more conventional forms, although this initial step will not render a completely familiar output for those readers who are experienced project managers. This phase revisits the objectives from a different perspective with the new goal being to identify the most appropriate and effective primary means to accomplish the objectives. Further, the primary means are further decomposed so that a set of fundamental actions are developed. For example, one of the primary means of accomplishing the objective centered on quality improvement in the manufacturing example is to improve the painting process.

While this is clearly one of the primary means to better quality, it does not define a fundamental action. To develop actions, consider improved painting as an objective and define the means to accomplish this. Two answers might be to use paint with a different chemical composition or increasing the frequency with which the painting equipment is cleaned. Each of these, now, represents actions that can be taken and are the level of decomposition demanded in this phase of the planning process. Typically, items that support the objective are called the primary means while the lower-level issues are called secondary means. This phase can be thought of as defining all of the necessary means (including fundamental actions) that are required to accomplish the objective. The remainder of this task creates a prioritization of these means and, in the process, begins to reveal existing synergy between the means in a way that the team members can both understand and exploit at a later time.

When this phase is accomplished, the planning process has made significant progress. A wide variety of ideas have been organized into objectives and the means to accomplish the objectives have been defined, carefully considered and prioritized. While this generic description of the process might sound rather conventional, two items will probably be encountered that are quite unconventional. The first is that several unique, and possibly conflicting, objectives will be defined and a satisfactory resolution is not obvious. The second is that it is quite possible that more than a hundred means will initially be defined as 'absolutely necessary' to accomplish the goals, yet they can't possibly all be accomplished.

4.2 TOOL: TREE DIAGRAM

4.2.1 Description

A TQM tool which is ideally suited to the iterative decomposition process needed in this phase is the tree diagram (sometimes called a systematic diagram or a dendrogram). Its structure is much like a family tree or organization chart with the trunk representing one of the 'objectives'. Actually, two sets of trees are constructed during this phase. The first set emanates from trunks that represent the objectives; the branches indicate the primary and secondary means necessary to accomplish the objective. The tree is grown in this fashion until all ideas are exhausted. It is desirable that growth terminate at a point where the nature of the branch titles corresponds to fundamental actions. It will be obvious that each level of the tree reveals more detailed information and the process naturally progresses towards fundamental action. It might be helpful to quickly scan the example at the end of the chapter and notice this progression in more concrete terms.

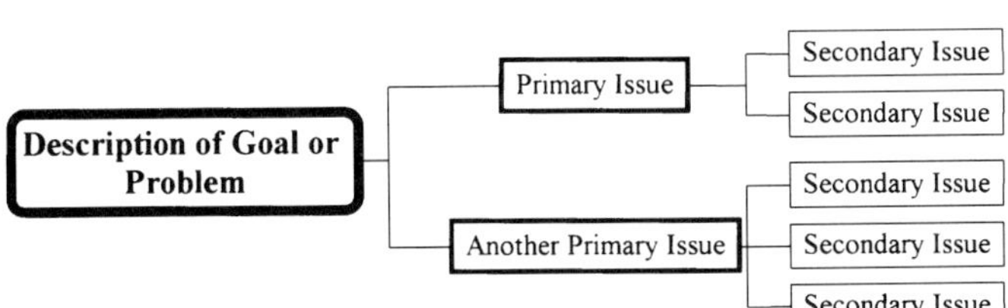

Figure 4.1 Example tree diagram

The second tree is constructed from results of the first, but rather than aiming for a list of fundamental actions, we now seek a prioritization of the means. Specifically, the means that were developed previously are now extracted from their setting in the first tree for rearrangement. Each goal forms the trunk of a new tree and the branches are developed by successively asking the question, 'What is the most important item necessary to accomplish the task?' and arranging the means that were created in the preceding exercise to answer this question. Once a set of branches are developed, the process is repeated for each of them. First-time readers might be confused concerning the differences between these two trees. The first tree is the result of decomposing an important idea into it parts regardless of its relative importance to the total success. Typically, this must be accomplished before it is easy to see which of the items must be performed first to create the greatest chance for project success. That is to say, the two trees will be different simply because the objectives are different.

4.2.2 Construction

Constructing the first tree diagram is not especially difficult; however, this section is designed to both provide some ideas on making the process as efficient as possible and enumerate a few common pitfalls. This task is performed by the entire team. As such, it is necessary to carefully record the work in such a fashion that the entire team can witness the development of the trees and so that a permanent record is made of the process. The basic steps in constructing the first tree are identified below.

1. The objectives that were identified during performance of the tasks indicated in Chapter 3 are identified as the trunk of a tree.
2. Working on each tree individually, first identify the primary means which are necessary to accomplish the objective. Then all secondary means are deduced with each level of the tree representing the necessary items to accomplish the immediately preceding level. Recall that all the issues below the

primary means are entitled secondary means, although there may well be a number of levels present. It makes no difference whether the procedure identifies all primary means first and then proceeds to the first level of secondary means etc., or whether one branch is fully developed until termination. In general, Mizuno suggests the following guidelines when developing the means:

(a) start with the principal means and go to the secondary ones through association;

(b) start with ideas of means considered least important and arrange these in groups, building up to more important ideas;

(c) acquire ideas of means without questioning their importance.

These recommendations can be implemented through the use of a brainstorming-type session with brief interludes to construct small affinity diagrams and much success has been made using this procedure.

3. When the fundamental actions have been developed, it is necessary to evaluate their practicality. An easy way to accomplish this goal is to use the following symbols beside each fundamental action: ● = practical, ○ = probably practical with a little work, Δ = uncertain practicality, × = impractical. The general idea is to make decisions on all fundamental actions and, as soon as possible, resolve the items identified as 'uncertain' to either of the other three categories. A word of caution and encouragement here is that ideas which are radically different will, on hasty consideration, most often been judged impractical. Before slipping into this trap, try to reinforce the idea with others in the list or take an approach which makes it practical before rejecting the idea. It is often the radical sounding idea from which the most important breakthrough develops.
4. The second tree is now developed from the first. The objectives and all of the means (including the fundamental actions) are written on either cards or the Post-It notes. As before, each objective forms the trunk of a separate tree. To avoid confusion, our trees are free to grow in any direction,

right to left, left to right or even upside down! The trees are grown by, first asking the question, 'Which of the means that we have identified is the most critical to accomplishing the objective?' If more than one is determined to be critical, place them side by side. The next level of the tree is developed by successively looking at the first level and asking, 'Which of the remaining means is most necessary to accomplish this one?' The cycle is repeated until each of the trees is fully developed. It is permissible both to add new ideas if they arise and to prune ideas that don't now seem appropriate. Finally, when all trees are developed, review their content to ensure that if all of the identified items are accomplished in the order presented, the objective will be met.

4.3 SUCCESS FACTOR – DEFINING SCOPE

The ultimate success of this phase is critically dependent on the quality of the objectives defined in the previous phase. While the techniques here provide several opportunities for internal correction and modification, they provide no check on the appropriateness of the input objectives. During the performance of this phase, it may be necessary to make adjustments to the objectives or revisit the tasks outlined in Chapter 3. The output of this phase should yield fundamental actions that are arranged in such a way that, philosophically, is consistent with each team member's intuition concerning tasks that are important to the success of this project. Since the project team represents a diverse set of views, it is equally likely that the sheer number of tasks looks daunting.

It is also important to recognize that the problem is being probed from a number of different dimensions that require different views. This is the stage where important breakthroughs will occur if they are at all. Encouraging radical ideas and nurturing them into practicality take patience and time. The easiest path is the 'do what has always worked', even though it may, in fact, not work too well. Any extra time which can be spent in defining the 'objectives' and the trees will yield tremendous dividends when the project is complete and the review is taking place.

The problem definition portion of TQM-based Project Planning is now complete. Although the work here clearly defined a path by which all 'objectives' can be met, it is painfully obvious that the number of critical actions identified cannot all be accomplished in their current form. Hence, it is time to move into system analysis and try and combine the tasks in a way to accomplish the most 'objectives' possible.

4.4 MANUFACTURING EXAMPLE

Pi and the group are becoming more comfortable with each other and the results of their efforts, although the process has been painfully slow. It has been two months since Pi interviewed the customers and three weeks since the team was chosen and informed of their task. The meeting today is dedicated to development of systematic diagrams that will identify the means to accomplishing the objectives. Recall, the objectives were to improve printing, improve manufacturing quality, explore different designs, develop creative marketing strategies and control costs. The systematic diagram is designed to define the most effective means for accomplishing the objectives. To keep our example tractable, we limit the example to the objective aimed at improving the perceived looks of the plane – specifically, improving the printing process.

After the meeting time was scheduled for the team to construct the tree diagram and the agenda sent to the members, Pi received an interesting note from the marketing representative and outside sales person. The note indicated that they would arrive towards the end of the session because their input wasn't required until the last two objectives (e.g. marketing and cost control) were to be discussed. 'Clearly', Pi thought, 'this process is not as obvious to everyone as I thought. I suppose it's time to add an agenda item to discuss the need for varying input on all topics; otherwise, we will just resolve this problem the same old way.'

The situation that Pi faces is not uncommon. As the topics begin to become more focused, members with expertise different from that which is obviously demanded by the subject may well begin to assume their counterparts will 'carry the ball'. While this reaction may be natural, TQM-based Project Planning depends on its not occurring. The planning process is still in a phase where it is incredibly important to capture input from a variety of sources. As you will see from the tree diagram, the marketing and sales people had substantial input. For example, looking at the printing components of only vaguely similar products was the outside salesperson's idea. It was captured during a comment that went something like this, 'Did you ever wonder how in the world McDonalds changed from using that wax-type paper to one that feels like regular paper but the printing never smears on your food? I wonder if it's the ink or the paper or the process or . . .' Your team will need encouragement to continue being active participants if this is to be successful.

The tree diagram for improved printing is illustrated in Figure 4.2a.

Notice that this application allows the tree to grow left to right, with each level deducing the means for the previous level. Specifically, the group decided that improved printing had three unique enabling processes: (1) expand the range of colors, (2) improve the print resolution on the final planes, and (3) make the ink more durable. Subsequently, each of these are considered 'objectives' and their means are deduced. The group, for example, decided that possible means that would allow the range of colors to expand included considering other suppliers, looking at different types of ink, mixing ink in-house, and considering using wild, neon colors rather than the traditional ones so encouraged by the corporate culture. Each of these, then, are considered the 'objective' and their means are deduced. This process continues until a viable list of fundamental actions are developed.

After the fundamental actions are defined, the next step is to assess their practicality as using the symbols ● = practical, ○ = probably practical with a little work, Δ = uncertain practicality,

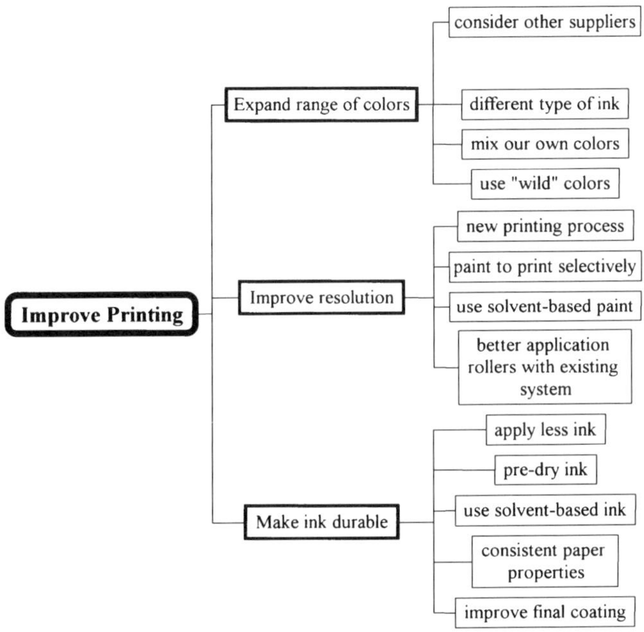

Figure 4.2a Tree diagram on improved printing

× = impractical. The results of their deliberations are illustrated in Figure 4.2b by the symbols to the right of the fundamental actions.

The group engaged in rousing discussion when some of these items were on the table. For example, the idea of developing an in-house ink producer was appealing to many in the group, but the investment and liability realities virtually eliminated any realistic chance that this approach could be chosen and it was determined that a much better expenditure of energy and effort would be on the other activities, all of which were given a better than average chance of success. The fundamental actions that required further investigation were divided among the team members for additional study and the meeting was adjourned. The action items for the next meeting were simple: (1) study the fundamental action that was characterized as either Δ and ○, (2) write a brief report on your findings, and (3) recommend that the action be reclassified as either a ● or ×.

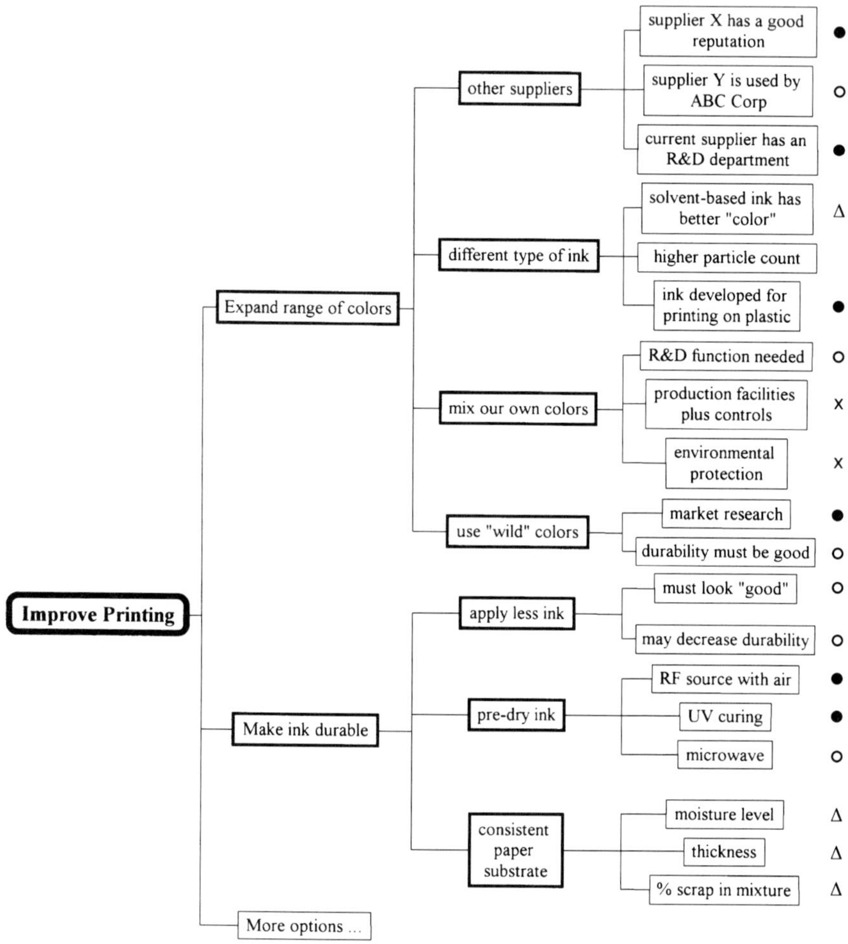

Figure 4.2b Expanded improved printing diagram

The team was reconvened the next week and made the reclassification. Some followed the suggestions made by the report writer and others, especially concerning solvent-based ink, were classified as practical, differing from the recommendation. The spirit of the discussion concerning this action centered on the consensus that solvent ink could resolve a lot of very difficult problems while only creating others that could be successfully handled with existing technology. That is, the group decided to label as practical a solution that created an entire set of problems – but ones that were thought to be easier to handle than the existing ones.

4.5 SERVICE EXAMPLE

Developing a tree diagram based on an affinity process helps us take advantage of the new opportunities the affinity diagram has provided. In this case, NR and the team see both similarities and differences between the affinity diagram and the present organization. Discussion of the affinity diagram develops three main points:

1. Configuration is primarily related to hardware and the function it performs, but has some aspect of software as well.
2. The Consultation area is primarily related to software, but has some aspects of hardware (especially in training) and networking.
3. The Networking area is very close to the present Network function provided by the computing support service function.

Further discussion brings out the idea of organizing computing support functions around hardware, software, and networks, as shown in Figure 4.3. With this in mind, a second round of brainstorming and discussion leads to new definitions of these function. Consultation, for example, becomes distributed across all three functions. The Hardware function, in keeping with a customer orientation of computing support, is related to hardware in the user's location. Software is divided between centralized and desktop applications, and Networks has been re-defined into logical groups that include the items from the affinity process.

Frequently, affinity processes are followed by tree diagrams. The affinity process is intended to bring out new structures, and the tree diagram is intended to modify, amplify and extend them

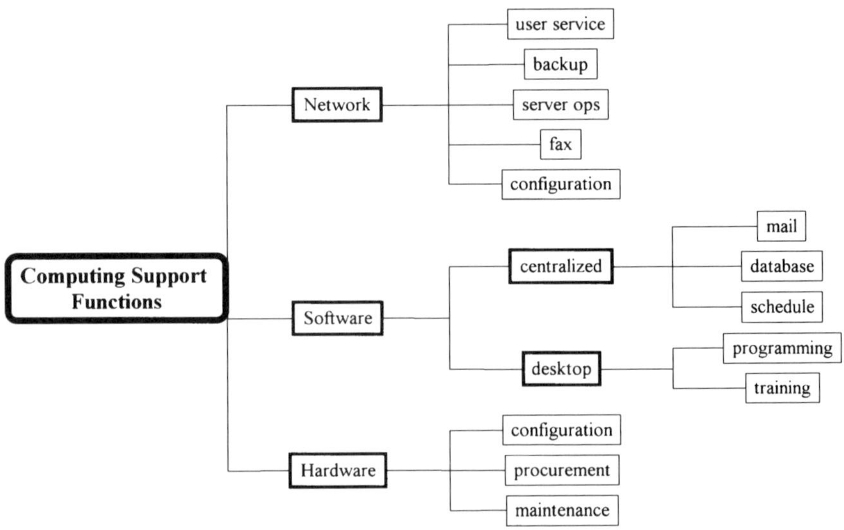

Figure 4.3 Tree diagram on computing support functions

with a higher degree of organization. The team has taken this point of view to arrive at a new organizational structure for computing support, based on their findings and observations as refined by the affinity process. A crucial point of discussion at this stage was whether the structure should be based on present functions or on anticipated new functions. The consensus in this case was that a drastic organizational change was not necessary, since the concensus structure was broad enough to accommodate expected future development.

4.6 SUMMARY OF KEY POINTS

- Build on the affinity grouping to find new informative structures.
- Add, remove or reclassify items as the quality of information improves.
- Don't be a slave to the tool; use it, modify it and annotate it to meet your needs.

PART THREE

System Analysis

INTRODUCTION

The system analysis phase of TQM-based Project Planning builds and synthesizes knowledge at a bit less conceptual level by molding the concepts and ideas developed thus far into a 'systems' format from which even more specific planning details can be constructed. You will recall that the fragments of information from the idea generation phase have been grouped according to their 'natural' affinity, but there are probably still a large number of items that feature very diverse concepts. It is also likely that some of the participants are now getting very uneasy because there is no obvious mechanism to transform this plethora of diverse concepts into a cohesive project plan. The solution lies in gaining system level knowledge and presenting the results in a way that selected details become more apparent.

To accomplish this, the following three tasks are required:

1. Causal relationships between the objectives are defined.
2. Flowcharts are constructed to illustrate these relationships and to begin identifying some of the constraints on the process. Further, the time required to complete each element of the plan is estimated, albeit crudely.
3. Knowledge of the system is developed and the specific actions required to complete the tasks are developed, albeit at a high, relatively imprecise level.

One effect of the first two tasks is that synergistic relationships among elements of the tasks are identified and consolidated. Another is that the most important sub-tasks are distinguished at a time when, if necessary, special provisions can be made to ensure success. Finally, the time estimates, although rough, provide insight into the lead times required for the critical tasks.

Clearly, we are beginning to make a transition from tasks and output not previously associated with project management to more familiar territory. On the other hand, the more familiar output is being constructed using tools and techniques that are different and the scope of the project will still appear untenable to some. In truth, there is a definite movement from the abstract and somewhat idealistic goals towards specific tasks and an ordering of those tasks. The details provided at the end of this phase will reinforce this fact.

As the reader proceeds through Part Three, it is a useful exercise to begin mentally to compare the key features of the system analysis phase with the project planning phase identified in the last chapter. As you will soon realize, system analysis tends to focus on:

- identifying causal relationships between objectives;
- providing a crude time-based structure to certain actions;
- gaining more specific 'system' knowledge while project definition is fundamentally concerned with:
  - obtaining the broadest and most diverse spectrum of creative thought and ideas;
  - discovering natural relationships among a number of basic ideas;
  - organizing the objectives according to importance.

Hence, you will find that while project definition deals with a tremendous number of ideas, all of which are treated as equally important, system analysis quickly begins to prioritize the objectives according to a number of criteria and begins to focus our immediate attention on a few critical factors. This is the first real indication of the 'broad to narrow' philosophy associated with TQM-based Project Planning.

Recall that the project definition phase dealt exclusively with ideas, concepts and goals that are internal to the project and were developed without any restrictions. In fact, pains were taken to ensure that ideas were promoted regardless of their ability to be implemented. Now, system analysis takes these wildly varying issues and begins to introduce external factors that have an impact on the project. Thinking about these broad concepts and gaining an intuitive familiarity with the 'big picture' will provide a much clearer understanding of the overall process and assist in avoiding some of the common problems of viewing TQM-based Project Planning as the progressive application of TQM tools that, if blindly but religiously followed, will produce the desired outcome.

The fact is that there are many tools that work equally well in accomplishing the specific tasks required in this methodology. Further, it is quite possible that your company has developed or purchased a proprietary methodology aimed at addressing one or more of these ideas that works better or is easier to implement than those tools presented in this text. If that is the case, please use the tool. The true value of TQM-based Project Planning is not in the recommendation of a particular TQM tool; rather it rests squarely in the methodology presented. Hence, one of the keys to success is a willingness by members of the project team to learn the methodology and stay focused on the tasks at hand.

CHAPTER FIVE

Key relationships

> To master and exert maximum control over this relationship, a company must be able to seek out and identify the real causes that lead to certain effects.
>
> (Alberto Galgano, *Companywide Quality Management*, 1994)

5.1 PLANNING OBJECTIVE: DESCRIPTION OF OBJECTIVES

Since the phases of TQM-based Project Planning are sequential in nature, the output of one phase is the input to the next. Thus, the input to this phase includes several systematic diagrams that represent the objectives from different perspectives and preliminary precedence relationships among them. The arrangement is hierarchical, a natural by-product of the methodology used for construction as well as the simple fact that tree diagrams lead to this arrangement. The objectives at the highest level, nearest to the tree's 'trunk', frequently represent those that are the most important and critical to project success while those further from the truck tend to be less critical. Ignoring reality for the moment, the tasks and actions identified in the tree diagrams should represent the elements for a tremendously successful project, if they could all be completed. In other words, accomplishing the entire set of tasks would ensure that all the objectives identified by the team would substantially be accomplished.

Clearly, there is little prospect of securing the infinite resources required to make this realization true, so priorities must be made. It is desirable that the limited resources available be expended such that the largest number of the most important objectives have the greatest chance of being accomplished. This phase moves our project plan from the 'pie-in-the-sky' towards a more realistic and prioritized set of tasks that can be accomplished in a limited amount of time.

There are frequently causal relationships between many of the objectives that must be identified and, subsequently, exploited if in this implementation we are to construct a plan that will truly deliver 'more bang for the buck'. The principal goals for this phase are: define and refine the known relationships, discover unknown relationships, and prioritize all relationships with respect to their importance in achieving as many objectives as possible. The technique proposed for use in this phase is a graphically based TQM tool that enhances the ability of the team to locate the important objectives and relationships from a large list. To obtain a conceptual understanding of the process, first visualize writing all the objectives on a very large piece of paper. Then, one by one, draw arrows from each objective to all other objectives with which this objective has any relationship, regardless of reason. Clearly, objectives that possess relationships with a large number of others will certainly be candidates for early completion. If the task has characteristics that prevent it from being quickly completed, then it must treated as a potential bottleneck and managed carefully so that it doesn't have an adverse effect on the remainder of the project. Understanding and quantifying these kinds of relationships are at the crux of constructing a plan that meets a large number of objectives.

Now, some of these relationships are obvious to everyone on the team and have been for a long time; so, a legitimate concern centers on the wisdom of devoting an entire chapter in this book, and the associated time in a meeting, to formalizing the obvious. The answer is multi-faceted. First, all relationships must be carefully, thoughtfully and comprehensively captured on paper whether they are obvious or not. Two common pitfalls associated with this are: (1) 'this is so important nobody could possible forget this one' . . . and everybody does and (2) that which is obvious to one person isn't to everyone in the same way. Second, the team will discover a number of previously unknown relationships during this process, some of which will be highly important to the success of the project. Many times, in fact, a few of these late revelations will seem so completely obvious and important that the team will wonder why they didn't know that in the first place. It is worthy to note the tools used force the team to look at issues from a variety of angles and this is the key to seeing previously obscure relationships very clearly. Regardless of how well known the relationship was prior to the exercise, all relationships must be treated consistently because the final step in describing the relationships is to prioritize them according to importance (e.g. which ones have the most others depending on them).

5.2 TOOL: INFLUENCE DIAGRAM (OR INTER-RELATIONSHIP DIGRAPH)

5.2.1 Description

The influence diagram is a very simple graphical tool aimed at bringing causal relationships among the objectives into focus. This is achieved by drawing arrows from each objective which is considered a 'cause' to all objectives which are the effects. The tree diagram serves as an excellent starting point; however, there are a few important differences that allow team members to resolve some points that were troubling during construction of the tree. First, constructing the tree diagram depends on agreeing on the single most important element that is necessary to achieve the objective under review. Contrast this with the current goal of identifying all causal relationships, not just the most important one.

Another difference is that relationships are sought between objectives that originated in any of the trees. Although the influence diagram can become large and rather complicated looking in this phase, the key inferences to be made are almost easier to make as the number of arrows increases. Specifically, the objectives of maximum interest are the ones that have the

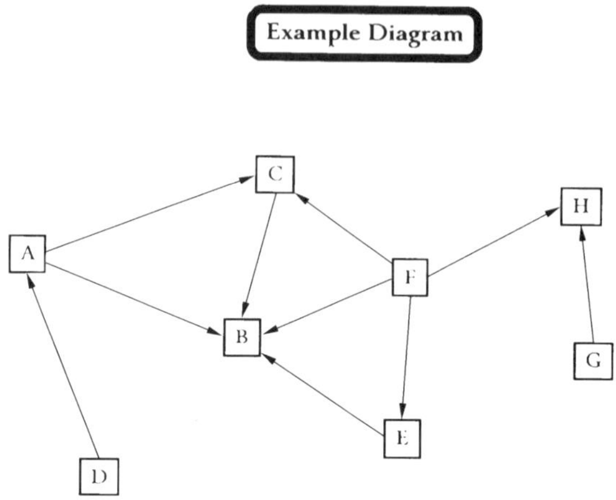

Figure 5.1 Example relationship diagram

largest number of arrows associated with them. Those that have all (or nearly all) of its arrows pointing outwards are important to accomplish early in the project because the success of many other objectives is dependent on an early and successful completion of tasks associated with this objective.

On the other hand, an objective that has a large number of arrows pointing towards it is probably a critical 'objective' to the ultimate success of the project because its success is dependent on success in a wide variety of other objectives. Those objectives that have a large number of arrows but no majority direction clearly possess features of both the previous types. These objectives are true enigmas because the outgoing arrows imply the objective must be completed as early as possible but the arrows pointing towards this objective ensure that the objective can't be completed until later in the project. In practice, these types of objectives require great care and must be carefully managed. The three types of objectives with a large number of arrows are special because they define the initial set of objectives associated with the starting point of the project (i.e. objectives with a majority of arrows pointing out), the ending tasks (i.e. objectives with a majority of arrows pointing in) and the objectives that will require the greatest amount of personal time and attention from the project manager (i.e. those with mixed direction).

5.2.2 Construction

Constructing influence diagrams in the context of TQM-based Project Planning is rather simple because much of the background work has already been accomplished. If the Project Definition phase had not been completed, development of the diagram requires a brainstorming session to identify the elements upon which the cause and effect relationships are to be established. In this case, however, discovering relationships between the already defined objectives is the goal. Given this starting point, the influence diagram can be constructed using the following steps.

1. Write the objectives on Post-It notes or cards. It is permissible to use the media from the Project Definition phase, although the team often has gained sufficient insight that rewriting the objectives provides the opportunity for some synthesis, analysis and rewording for added clarity and removal of ambiguity and/or duplication. Place the cards on a large piece of paper.
2. Systematically work through the objectives and draw arrows from each 'cause' to each 'effect'. One advantage of using the Post-It notes is that the team can use their intuition to initially arranged the objectives in a way that keeps the arrows more organized.
3. After all objectives are visited and the relationship arrows drawn, the influence diagram is comprehensively reviewed for accuracy and consistency. It is permissible to add objectives if the team feel they are necessary or to combine some of the objectives if appropriate. The goal is to develop a comprehensive and believable diagram which will form the basis for the project flowchart. Spending sufficient time at this task to ensure that the team feels the output is usable is essential for success of the remaining phases.

5.3 SUCCESS FACTOR – MAINTAINING FOCUS

A key to completing the influence diagram successfully is maintaining focus throughout the process in spite of size of the problem. Team members must understand that a properly constructed influence diagram is a critical pivot point in TQM-based Project Planning. This is the first place where the somewhat amorphous and abstract objectives begin to be shaped into a prioritized, chronological list of tasks that must be accomplished. While the output is only a first-cut at a project plan, it really establishes the basis for the final set of tasks. Further, identifying those objectives with a larger number of arrows in the influence diagram points towards critical components of the plan as well as potential bottlenecks. Planning for the activities associated with these objectives rather than reacting to them during the performance of the tasks may prove the difference between ultimate success and failure of the project. Clearly, actually constructing the diagram is not terribly complicated, but maintaining focus and expending sufficient time and effort to synthesize and analyze the information is important for the facilitator to recognize.

5.4 MANUFACTURING EXAMPLE

As the project drags on through what some feel are interminable brainstorming sessions, Pi is beginning to feel the heat. Some of the technical experts are particularly obnoxious because, as their recent memos indicate, they already know everything there is to know about a particular area and all of the 'wasted' time in meetings could have been avoided if Pi had just asked them directly and implemented their suggestions. Even worse is her implication that anyone without 25 years experience in making paper airplanes could possibly understand the problems, let alone be capable of suggesting the answers – a testimonial to the incompetence of both her and 'upper management'. The *coup de grace* is the assertion that this planning exercise is 'analysis paralysis' and if the company (whoever that is) would just 'do something' then everything would be fine.

Despite this barrage of memos, Pi continues to push forward because she understands that, at this point, the project definition process has been somewhat utopian since the emphasis has been on inclusivity and wild ideas. Further, she realizes that the upcoming System Analysis phase (that begins with this chapter) begins to impose real-world constraints on the ideas and concepts generated thus far and some of her detractors will feel better as soon as they can veto some ideas. Further, any innovation will be greeted with skepticism, and the amount of time already taken to get to this point – a point that doesn't look like a traditional project plan – is subject to even more skepticism.

The actions outlined in this chapter seek to establish priority relationships in all places where they exist. Figure 5.2 illustrates the outcome of this process that Pi and the team developed for some of the objectives and fundamental actions associated with the printing issue. As indicated in this chapter, the individual items were placed on Post-It notes and arranged a few times before the arrows were drawn. After the arrows were constructed, the entire diagram was studied and discussed. The portion illustrated in Figure 5.2 revealed several interesting aspects.

First, the direction of some arrows depended on discussion, so the interpretation phase required recollection of the earlier points. For example, the idea of using solvent-based ink to improve resolution involved a lot of arrows. The reason is that there are a lot of safety issues associated with its feasibility, but this alternative also has the potential to resolve a number of printing-related issues.

The items with the largest number of arrows are typically very important to the technical success of the project. Other interesting features are those boxes which only have arrows either pointing towards or away from them. In this case, the new technology is one of the ideas that seem to be the end of the road. Pi and team weren't at all surprised by this result because they all realized that upper management would probably entertain a

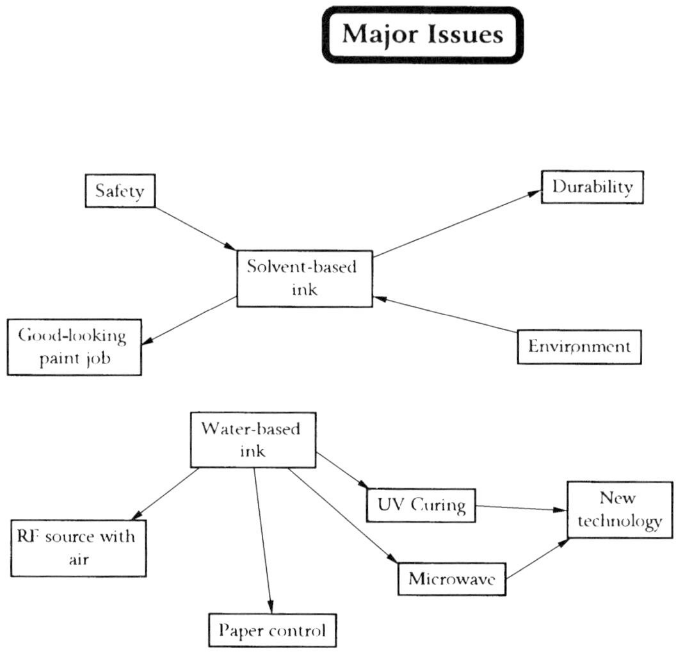

Figure 5.2 Major issues in ink type

capital request for this amount only if all other alternatives had been investigated, and then it might not be approved. On the other hand, the boxes that exclusively have arrows pointing away are all investigatory in nature and will help define the pathways that seem most likely to succeed and identify the roadblocks to their implementation.

These results appealed to the team's collective common sense and immediately perked the enthusiasm for moving onward. On the other hand, the results of performing this exercise for all of the objective allowed the members to gain insight that was not at all obvious. For example, the complete exercise reinforced that the most recent advances in solvent-based ink were critical to other higher level objectives as well. Another discovery was made after several members mentioned that the level of discussion and the number of arrows leaving the fundamental action on capital justification seemed odd for an established company because it is well known that this is a requirement for spending capital in all companies. It was concluded that the reason for both of these centered on the fact that there was not a well-defined and well-known prescription for constructing a capital request. Upon further discussion, capital request was being made in the same format that the founder and her partner had developed when there were a handful of employees and making payroll was, at times, exciting. The inclusion of accountants, then managers and, finally, layers of management had increased the demands for better information but everyone had their own idea of what was needed. As a result, Pi requested that the Manager of Accounting develop a universally acceptable process for capital appropriations. It was in place within a month.

5.5 SERVICE EXAMPLE

One of the areas mentioned most frequently in the team's data collection was the anticipated need to expand network support. It was mentioned in almost every group consulted, but little detail was provided in the initial round. The team members agree that this is one of the areas for expansion, but they are also concerned that this expansion has so many alternative approaches that it could become unmanageable.

NR has decided to take the team through the development of an influence diagram to better define the issues in network expansion. This begins with brainstorming based on the question 'What are the key issues in expanding network support?' This leads to the diagram in Figure 5.3, with the typical process of placing new items on the chart and drawing arrows between them and all existing items as they are placed.

The team's analysis of the diagram begins with an arrow count, and the item with the largest number of arrows is 'Needs Assessment'. Clearly, the needs to be satisfied by network expansion must be determined before substantial design work can begin. The team also notes that two items have only out-going arrows. Of these, the 'Desktop Inventory' appears to

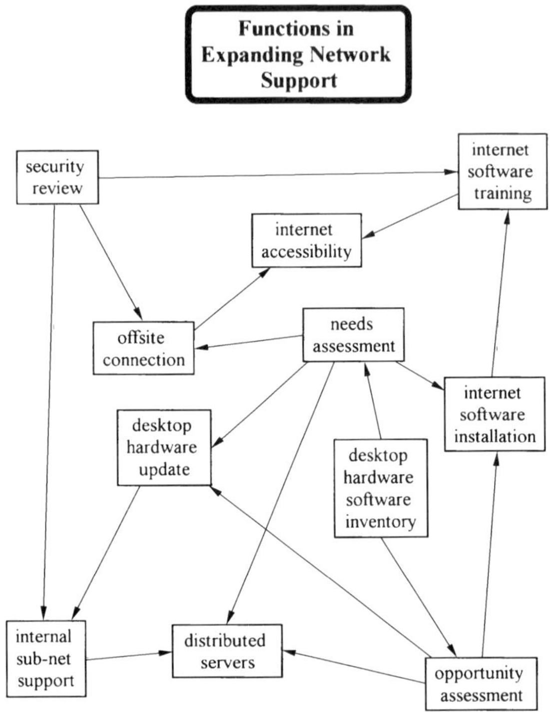

Figure 5.3 Functions in expanding network support

precede the needs assessment; this then defines an activity that must have priority. The security review must also precede several others, but it will not hold up other priority activities. Items that have only in-pointing arrows obviously must come later.

This analysis is typical. Rather than hold to a strict set of rules on numbers of arrows and their directions, the diagram is interpreted subjectively, and is used as a guide for further analysis. As with many of these tools, their importance lies in their graphical depiction of information and relationships. Their value comes with the discussion and decision making that they support.

5.6 SUMMARY OF KEY POINTS

- Focus on the relationships rather than the details of activities.
- Keep in mind the concept of the 'key log', that can create or release a log jam.
- Consider relationships other than sequence or precedence.

CHAPTER SIX

System structure

Flowcharts can be applied to anything from the travels of an invoice or the flow of materials, to the steps in making a sale or servicing a product.
(Michael Brassard and Diane Ritter, *The Memory Jogger II*, 1994)

6.1 PLANNING OBJECTIVE: DEFINING OVERALL STRUCTURE

Developing a common model of the project is the next objective in TQM-based Project Planning. At this point, the team has developed not only several tree diagrams, each featuring a major theme in the project, but also the influence diagram from the previous phase. While it would be ideal if there were but one influence diagram, reality indicates that several diagrams were developed. That is, even though all objectives were possibly placed on one sheet of paper, there are several sets of objectives that have arrows joining items within the set but there exist few, if any, arrows between the sets of objectives. The very nature of the tasks until this point encouraged the team to make groupings of tasks according to those with maximum affinity. Now, attention is turned to development of a common model which fuses the objectives into one project and begins to specify precedence relationships and durations for accomplishment. The general structure is reminiscent of the traditional PERT structure.

In fact, two flowcharts will be created during this phase. The first flowchart reflects a 'macro-level' view of the project. The goal is to organize the key objectives with respect to the precedence relations that have already been identified and to fill in the gaps with other relatively less important objectives until the components are sufficiently cohesive that all members of the team have a general, common notion of the overall progression. If this is viewed as the skeleton of the project flow, then the second flowchart begins to add just a little flesh to the bones. It is, indeed, a more detailed expansion of the first flowchart and seeks to achieve more consensus concerning the relative importance of the key objectives, the objectives of lesser importance per the influence diagram but critical to a cohesive flow, and progression between (and among) these objectives.

Clearly, the goal of gaining any type of consensus at this point may appear quite ambitious, but some very important hurdles have already been cleared. First, considerable effort was expended in the previous task constructing the influence diagram which not only defined the key relationships but also initiated a discussion of precedence and importance of the various objectives. The idea is to use the key objectives and relationships developed in the last phase as the basis for the macro-level flowchart. Then, use these key factors along with the 'subordinate' objectives and their relationships (both to each other and to the key objectives) to develop a general progression that carries the project from start to finish. This project is not intended to look like a finished product nor does it have to have an absolutely continuous progression, just a general flow that makes sense. From this point onward, this basic model will be reformulated and fined to achieve a seamless transition between correct tasks.

6.2 TOOL: FLOWCHARTS

The tool used in this section is hardly one associated with new 'Total Quality' concepts. Flowcharts have been widely used in industry and academia for decades and, unlike the other phases of TQM-based Project Planning, the difficulty here is not associated with learning another tool but properly applying an old friend.

A key to success will be continually resisting the urge to develop more detail than is desirable and demanding that each transition point is continuous and understandable. Another way to visualize the appropriate level of development in this phase is that the overall flow should appeal to intuition on a conceptual level, but the details making transitions between individual tasks or between different themes will not be developed and the chart will not withstand the scrutiny of detailed, implementation-oriented questions. Further, even though the second flowchart is designed to add detail, the amount of additional information will not satisfy those team members seeking definition of the day-to-day tasks that must be monitored and controlled for the project to be successful. In the overall scheme, these flowcharts are still at a conceptual level that has been labeled 'macro-level', and the meaning of 'detail' is relative. If we rate detail encountered in the brainstorming as a 1 and the detailed, daily project plan as a 10 this element moves us towards a 3 or 4, depending on the project. On the other hand, the movement towards more specificity is clear and unambiguous.

6.2.1 Top-down flowchart

The first task is to organize the key relationships into an initial model of the project. Typical flowcharts place these key objectives in blocks and, then, connect the blocks with arrows in serial and parallel paths to define a logical sequence through the project. Recall, it was noted during the analysis of the influence diagram that any objectives which had all arrows either pointing towards or from it was highly important and must be considered first during this phase. Now is the time to do this. The objective is to establish the most logical sequence of accomplishing these objectives and begin thinking of the precedence relationships among the objectives as well as the time required to accomplish each.

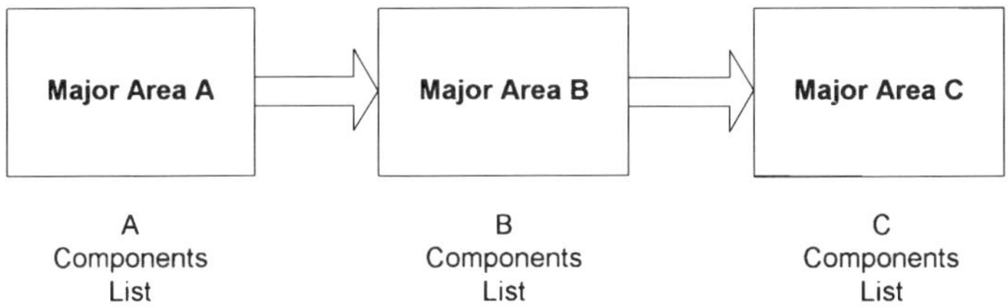

Figure 6.1 Example top-down flowchart

Use of cards or the Post-It notes is an excellent mechanism for constructing the flowcharts because there will probably be several flowcharts constructed before the one that satisfies the entire team is discovered.

After the key objectives are tentatively settled, it is time to add any objectives or tasks that are required to make the flow from start to finish reasonably smooth. The general idea of this path should be common to all team members, although a certain vagueness and ambiguity will certainly persist to varying degrees, since the issues which are being flowcharted are still rather global. The influence diagram can again be consulted to identify objectives which have a large number, but not all, of their arrows either pointing towards or away from them. These secondary objectives either are necessary on the macro-level flowchart or can spark the team into identifying a critical area.

6.2.2 Flow diagram

The second flowchart is the flow diagram which expands the top-down flowchart with some details. It is at this point that the temptation to jump into task-level details will be strong. It is important to resist this craving and remain focused on the objectives of this phase, developing the project components. The flow diagram seeks to provide a more comprehensive model of the project by adding secondary objectives and large tasks to the top-down flowchart. The team should be forming a more uniform and clearer picture of the flow from completing initial objectives through success in the ultimate project goals.

As the paths are formed, there are two additional tasks which

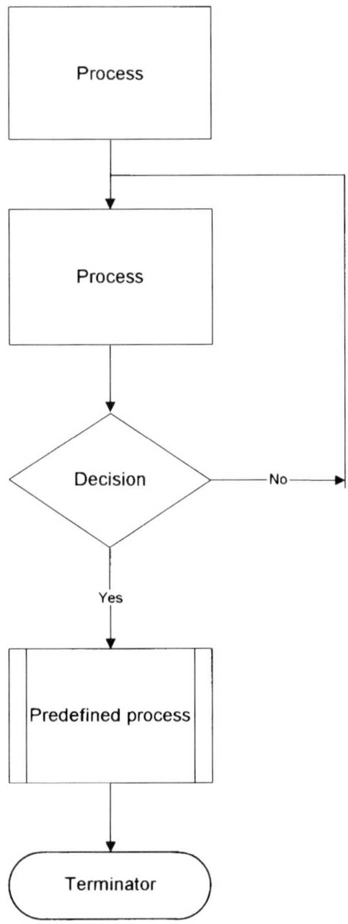

Figure 6.2 Example flow diagram

must be accomplished before this flowchart is complete: (1) the time duration which is required to accomplish the objective must be tentatively identified, and (2) the form of the flowchart must be analyzed. The first of these tasks is designed at forcing the team to begin to gain a sense of the scope of the project. The precedence relationships will add a dimension to the project that the team had not previously seen and the time element will add another. The impression of the project model should be that both length and depth are being added. The second task is aimed at taking an overall view of the paths which have been constructed to determine if the project has been described as either one long serial process or a very short effort but with a large number of simultaneous tasks. Either of these extremes reflect that some additional work must be performed. The desirable project flow will have a reasonable number of parallel tasks when this can be used as leverage in shortening the overall project time and serial tasks when required.

Construction of this flowchart can be accomplished using the notes or Post-Its as before. The added detail will normally require a bit longer to construct this chart because some of the nagging questions of team members will begin to be asked.

6.3 SUCCESS FACTOR – MAINTAINING A SYSTEM VIEW

As mentioned earlier, it may be very difficult to resist jumping into details of the project during this phase. Many experienced planners will feel that an uncomfortably large amount of time is being spent with the objectives. Rest assured that the 'nuts and bolts' will be carefully and fully addressed in due time and that dealing with the objectives in this fashion is designed to identify bottlenecks and critical objectives at a time when adequate resources can be devoted to ensure project success. Those phases of traditional projects should be avoided when the PM stays on the telephone trying to expedite some task or garnering short-term resources in a panic mode to facilitate completion. Now is the time to maintain the broader, systems viewpoint and deal with the objectives at the macro-level, even when developing the flow diagram that begins to identify some 'details'.

It is important to remember that this is a first draft of the project from start to finish. While the milestones are in terms of objectives and the techniques to attain these objectives are not defined, these flowcharts are the first diagrams which present a path from the current state, through a time ordered sequence of objectives to project completion. It is also the first time that the project has both length and depth and the team members should

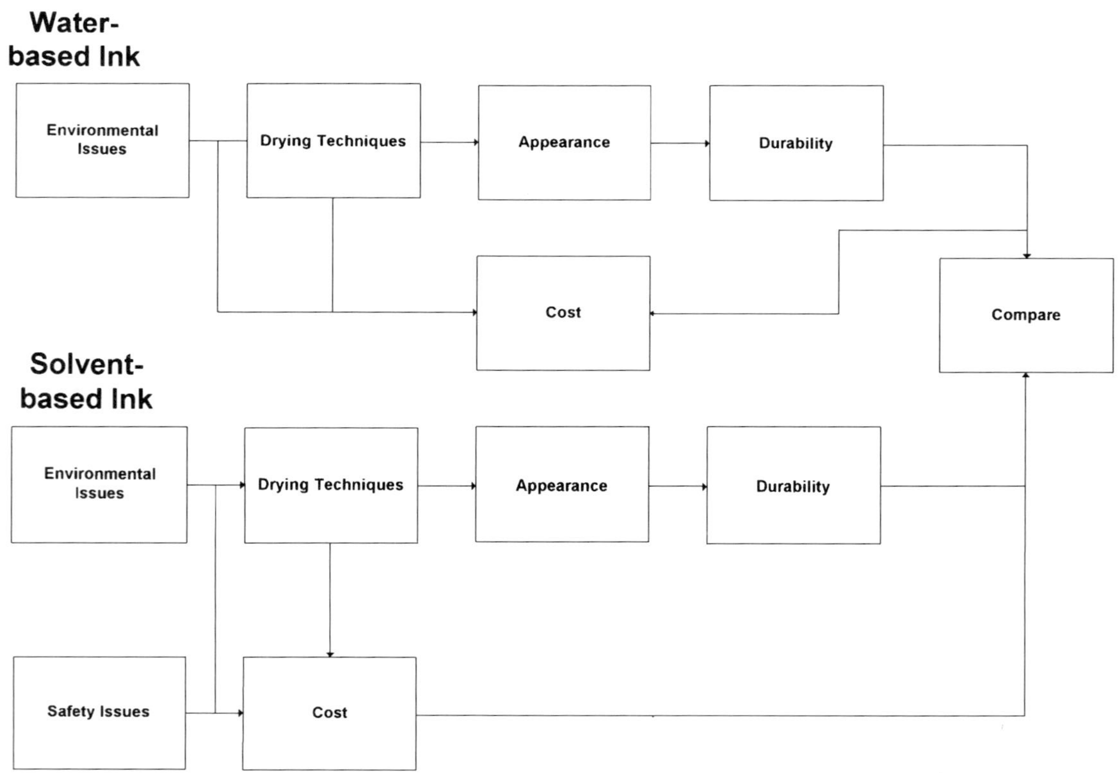

Figure 6.3 Top-down flow: water- and solvent-based ink

be developing a common sense of the sequence and organization. If this unanimity is not present, the flowchart should be revisited to clarify its content because lack of agreement here will only compound as more details are added.

6.4 MANUFACTURING EXAMPLE

As Pi and the team move towards more traditional-looking items, the number of temptations to 'do it just like we always have' become more abundant. In this task, the objectives developed in Chapters 3 and 4 are used with the precedence relations identified in Chapter 5 to construct a top-down flowchart that is, then, refined into a more detailed flowchart with rough time intervals associated with attaining each objective. Pi spends extra effort during the beginning of this session to ensure that the distinction between this exercise and creating a traditional PERT chart is well understood. Recall, the important difference is that this activity deals with objectives and higher level issues while PERT, as well as other traditional project management tools, typically deal with individual actions. After fully discussing this important concept, the team assembled the first flowchart, a portion of which is illustrated in Figure 6.3.

This figure illustrates a variety of activities that have been characterized as objectives. After careful consideration and a lot of rearranging the items on the chart, the team decided that the results were very satisfying to their intuition of the problem. Some of the somewhat easy go/no go objectives are accomplished early in the process while the more integrated aspects tend towards the later dates. Upon looking at the overall flowchart, decision on the mix of new technology, upgrading existing machines and living with old machines was important to other areas of the project, and these objectives were all accomplished rather early in the total project. The next flowchart focuses on adding details to the items and making crude estimates of the time required for each. Figure 6.4 displays the results for some of the segments associated with the equipment

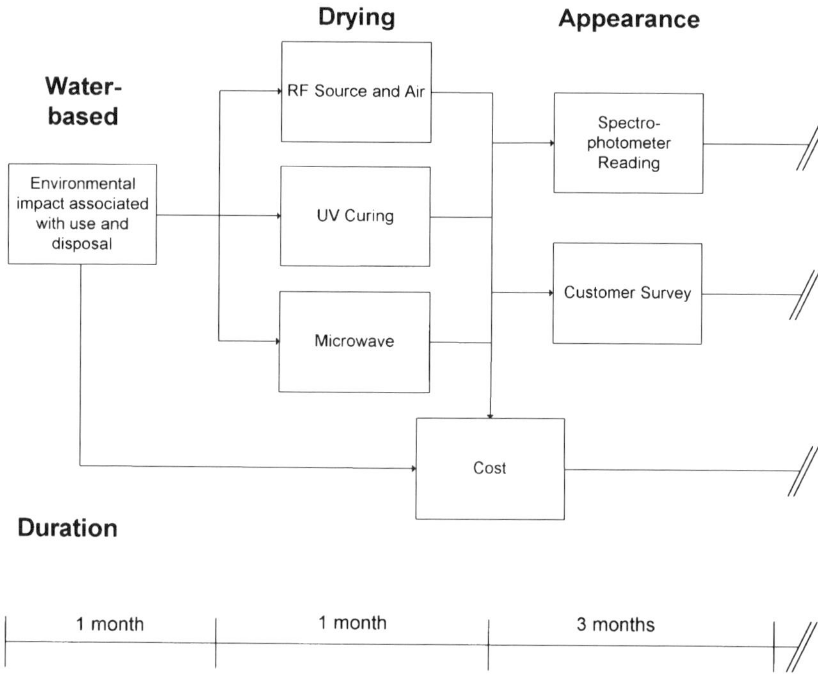

Figure 6.4 Process flow: water-based ink development

decisions. Looking carefully at these results, it is obvious that this flowchart still does not identify specific tasks that must be accomplished, rather, it identifies more precise objectives like 'develop a risk prevention strategy' without decomposing this into format, tool, system etc.

At the end of this exercise, the team was pretty tired but very satisfied with the results. Those, like Pi, who had faith in the new TQM-based Project Planning felt that the process had worked very well, but they were actually glad to see something that looked 'do-able'. The team members who had been skeptical of the approach felt vindicated that the results were beginning to look familiar, but they also saw many items on the flowchart that would never have been considered using the old methods – and the some of these had tremendous potential.

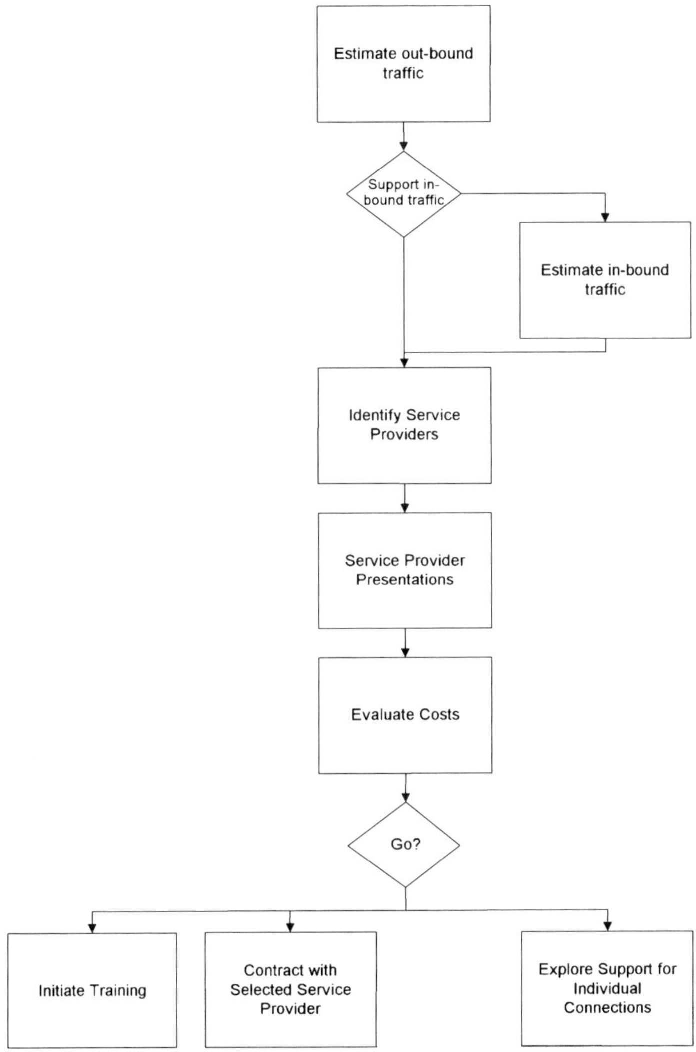

Figure 6.5 Developing external network connection

6.5 SERVICE EXAMPLE

As work proceeds on activities initiated from the influence diagram, new tools come into use. Several boxes in this diagram were related to connecting the firm to the Internet. A small sub-team has been tasked with further defining the issues regarding this, since the firm has little experience in this area.

NR and the team decide to construct a flow diagram of the decision process regarding Internet connections (Figure 6.5). This flow diagram should allow for flexibility in decision, and should logically present arriving at these decisions. Each box in the flow diagram can represent significant additional activity. For example the first box, 'Estimate Out-Bound Traffic', will require some assessment of demand, and could be part of the needs assessment. The decision block on in-bound traffic could depend on security concerns, and should follow this activity from the influence diagram.

Sometimes flow diagrams represent well-defined processes that should be followed repeatedly. In this case, the flow diagram is another decision tool; it represents a process that will probably be done once. In doing so, it provides a guide to the logical development of information that supports a sequence of decisions. The team will use it as they collect additional information and arrive at their final recommendation.

6.6 SUMMARY OF KEY POINTS

- Consider flow at two levels – macro and detail.
- Keeping the levels apart makes it easier to review and analyze them.
- Working down from the top, add detail where necessary to avoid confusion.
- These are good tools for documentation of standard processes.

CHAPTER SEVEN

System knowledge

Any rational plan, however simple, is prediction concerning conditions, behavior, performance of people, procedures, equipment, or materials.

(W. Edwards Deming, *The New Economics*, 1994)

7.1 PLANNING OBJECTIVE: DESCRIPTION OF THE SYSTEM

The System Analysis portion of TQM-based Project Planning is completed by focusing attention more fully on some constraints viewed at the system level. The project planning work until this point has dealt exclusively with the project objectives and goals. They have been developed, refined and analyzed from different angles to form a complete, macro-level, flowchart of the entire project. The activities in this task relate to the actual processes that will be utilized to achieve the goals of this project. Where no such systems exist or completely new technologies are to be developed, the focus will be on those components of the system which are known.

Philosophically, this information will, later, be integrated with the project flowcharts created in the previous phase to form the rather comprehensive knowledge base that can be used to develop a detailed, task oriented project management flowchart. For now, however, the goal is to describe the existing systems using information that team members consider common knowledge. In fact, the team will discover that little knowledge is common across the entire organization so this task is actually a pooling of information that exists within pockets of the organization to form comprehensive knowledge base about processes that will have an impact on the project.

Typically, two very positive tangential effects are also obtained. The first is that the team discovers that little, if any, knowledge is really common across the entire organization. This realization frequently results in more communication – informal, in most cases – between individuals. The other effect is that some of the stereotypes are dissolved between the functional areas of design engineering, manufacturing and manufacturing engineering, sales and management. The engineers rarely leave this process without gaining new insights and a sizable appreciation for the important of marketing in the company's success and the marketers discover that engineers are not exclusively calculator toting geeks whose sole purpose in life is finding numbers to charge their time against at the end of the week.

7.2 TOOLS: CAUSE AND EFFECT, PARETO AND MATRIX DIAGRAMS

The tools that are suggested to be used in this phase of the project are neither new nor unknown to most readers. Instead, they are included in the venerable 'seven old tools' or the 'magnificent seven'. Fortunately, these techniques are akin to wine rather than fish – they do, indeed, get better with age.

7.2.1 Cause and effect diagram

7.2.1.1 Description

The first tool is the cause and effect diagram (Figure 7.1). Traditionally, this diagram is used to facilitate diagnosing the possible root causes of a problem; however, this application will use its power to uncover cause and effect relationships in the system, whether the end result is a problem or a desirable outcome. The diagram seeks to define these relationships by initially defining the outcome to be analyzed and then channeling ideas in to six distinct categories, namely, man, machines, materials, methods, measurements and environment. A glance at the figure in the manufacturing example illustrates why this tool is also called a fishbone diagram. Closer examination indicates that the desired effect or outcome is the 'head' while the six broad categories (5 Ms and an E) of causes form major 'bones'. The remainder of the 'skeleton' is developed by reducing the causes into more elemental components. As the team places various objectives at the head of the fish and constructs the skeleton of causes, quite a lot of system knowledge is developed in an easily understood format.

The overall objective here is to look at the key objectives of the project with respect to existing processes that will have an effect. That is, we want to take the critical objectives that were defined in the previous phase and see how they are impacted by the existing physical and infrastructure systems of the organization. There is no single template that can be applied to all situations that will be encountered in your project other than the basic cause and effect structure; however, the examples at the end of this chapter illustrate typical applications. It is wise to remind yourself that until this task, real constraints have basically been ignored.

7.2.1.2 Construction

The diagram is constructed by first identifying the goal (the attribute or problem to be investigated) and, then, proceed to relate those items which have a causal relation on the goal. In this application, adding the details to the skeleton involves the using the project planning team and supplementing their knowledge with technical experts. These experts can provide invaluable assistance in developing sequences that involve a large number of technical issues that will not only provide a more accurate chart but save the substantial time associated with doing it wrong, possibly using this incorrect information later, and then re-doing it correctly. For example, suppose the goal is to deduce causes that make a piece of machinery automatically trip off line. The caveat is to ensure that everyone understands

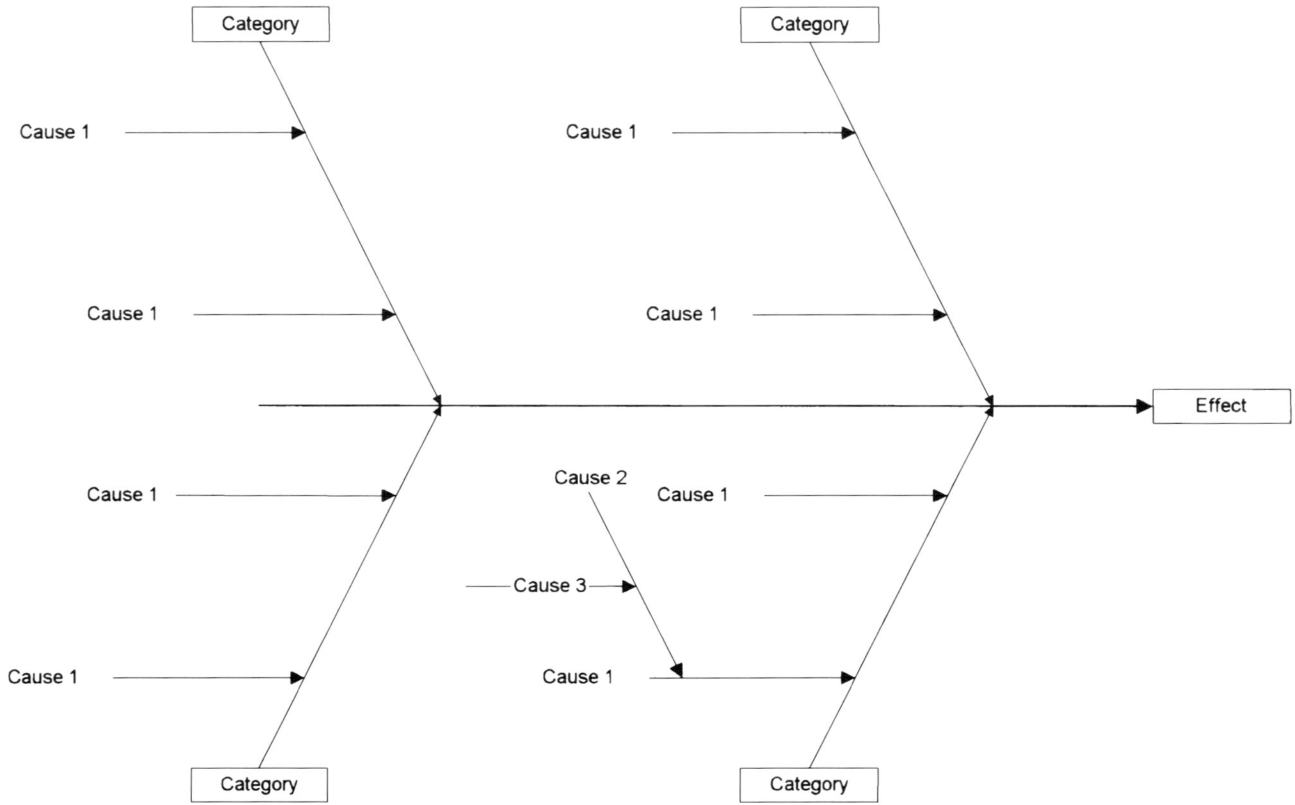

Figure 7.1 Example cause and effect diagram

that the experts are a resource. They simply cannot be allowed to dominate the direction of the project. One way to avoid this problem is to run this session as a brainstorming meeting, where all ideas are developed and the prioritization and pruning occurs later. Sometimes, different sounding ideas become useless but other times they are polished into the gems of the diagram. The format of the fishbone diagram has an element of focusing attention into the six categories, so the results are rarely as wildly divergent as a full brainstorming session.

7.2.2 Pareto diagram

In the eighteenth century, the Italian economist Vilfredo Pareto reported that 80% of the wealth in his country was held by 20% of the families. While Pareto's observation cannot be proven by traditional means, its proclivity to explain a wide variety of situations is certainly impressive and it commonly explains situations in a wide variety of circumstances. You were probably introduced to it as 'Pareto's Principle', with an implied certainty of $F = ma$ and $E = mc^2$. In TQM-based Project Planning, this concept not only provides relief for the team but is somewhat of a cornerstone for making TQM-based Project Planning viable in practice. There is little doubt that the team members and the system experts will define a large number of parameters during the system definition phase and while the task of monitoring and controlling all of these items would be nearly insurmountable, addressing only 20% of them is frequently thought to be reasonable and achievable. Project planners and those interested in system improvement take solace in the fact that the Pareto Principle can be thought of as implying that correctly addressing 20% of the system parameters will provide control over 80% of the problems. Finding the right 20%, however, is an important issue.

The Pareto diagram (Figure 7.2) is highly useful in this context. It is used to display the types of defects (alternately, types of problems) on the x-axis and the fraction of the total number defects (or problems) that particular type represents on the y-axis. For example, the vast majority of all projects are initiated

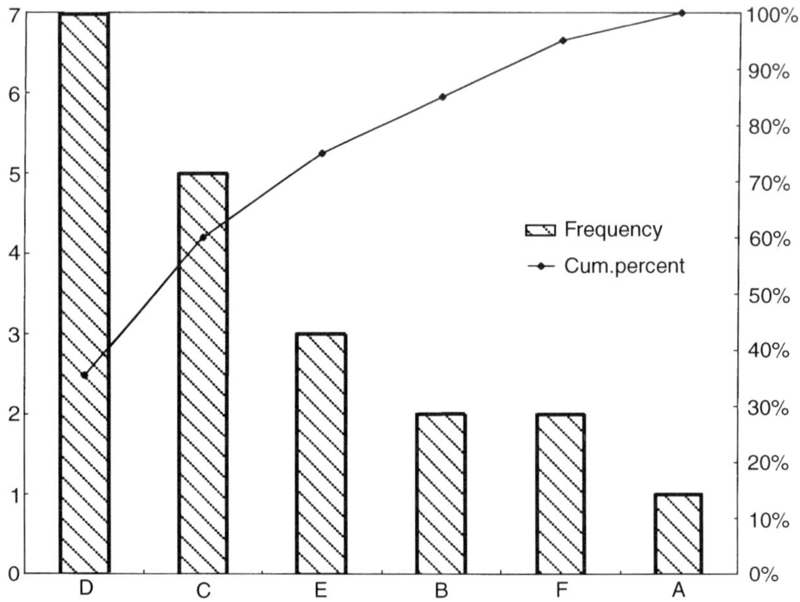

Figure 7.2 Example Pareto chart

as the result of customer feedback and the results of most projects are an incremental improvement over an existing product or service. (It is extremely rare to be tasked with designing a new generation of project that is not at all similar to the existing one and will not use any of the existing processes for production.) These facts indicate that there will likely be existing processes as well as data on a product or service very similar to the focus of your project. This data should be analyzed using a Pareto diagram. Using this technique, the team can focus on the most critical problems. In project planning, especially when an existing system or systems are to be modified, it is often also instructive to focus on the portions of the system which provide the fewest problems. This frequently offers some insight into things that work well with the particular type of system at hand and, further, helps identify some of the critical systems and parameters to be controlled.

7.2.3 Matrix diagram

Matrix diagrams are used to assist team members focus on the most important parameters. While matrices have historically been solely associated with mathematics and quantitative thinking, a number of recent publications indicate that the basic matrix structure is frequently used to make qualitative assessments of items. In a simple two dimensions array, the first set of factors ($x_1, x_2, \ldots, x_m$) are arranged along the left side of the matrix forming rows and the second set of factors ($y_1, y_2, \ldots, y_n$) are arranged across the top of the matrix forming columns. The team then fills in the matrix using different icons to denote the varying strength of the relationship. In the example (Figure 7.3), ⊙, ○ and △ mean strong, moderate and weak relationships respectively. This approach helps to identify important factors (e.g. those associated with a row or column containing a large number of icons indicating a high degree of importance) and helps build a basis for further analysis. Although this example is limited to two dimensions, there are other matrix formats allowing higher dimensional relationships to be seen.

Clearly, matrices are not limited to presenting an array of numbers for clever mathematicians. They are very potent tools for identifying the important factors from a large set, exactly the task that needs to be performed at this stage of TQM-based Project Planning. The uses of these matrices is limited only by the imagination, and a perusal of the total quality literature can provide insight into a wide variety of matrix templates for use in different situations. The manufacturing example illustrates the use of matrices for depicting qualitative data.

	Item 1	Item 2	Item 3	Item 4
Factor A	○	⊙		
Factor B		○		△
Factor C	△		⊙	

Figure 7.3 Example matrix diagram

7.3 SUCCESS FACTOR: MANAGING ADDED DETAIL

This phase of the planning effort requires considerable caution to avoid becoming overwhelmed by the added detail that is being developed. While it is indeed true that, in theory, one could identify much of the detail and, then, pare down to the few important elements, this technique rarely is successful. This technique, however, too often leaves the team members frustrated and despondent when they view the sheer amount of information to be processed. Further, the results are rarely as good or the process as efficient when compared with the recommended approach that carefully manages the added detail because the additional information is gradually added in a very focused manner. In addition, the new material is synthesized and woven into the existing data to create a coherent story that grows in size and detail, but remains understandable during the interim because deviations that confuse the real issues are detected and modified before they are integrated. The tools identified earlier in this chapter: Pareto, cause and effect, and matrix diagrams, are ideally suited to assist the team in describing the system in this fashion. Don't forget that the objective of this phase is to paint a more complete picture project by imposing both realistic constraints and prioritizing the issues according to historical data. This can only be accomplished by using the wisdom and insight of a variety of people. As team members and selected experts begin to provide input, it will be worth the effort to become familiar with these tools so that the information can be correctly handled. You want to include all of this knowledge, but at the time when it can be properly managed and used.

7.3.1 Focus with Pareto and cause and effect

The initial challenge associated with this phase is to create areas upon which to focus during the knowledge-building exercise. The Pareto diagram can assist the team in organizing any customer feedback or other data on similar products (alternately, services) or processes expected to be used in this project in such a way that the important elements are identified. Industry has historically used the Pareto diagram to classify problems (e.g. types of defects or customer complaints) so that process improvement efforts can be targeted at the areas of greatest need; however, the use of these diagrams should not necessarily be restricted to this application. If the Pareto principle is true for your application as it is in many, nearly any data of appropriate form can be charted and the important items are easily distinguished. If the project is improvement oriented, it might be useful to solicit and chart customer comments, both positive and negative. The positive comments will point to areas of the process which are very good and need to be preserved while the negative ones clearly define areas in need of improvement.

On the other hand, constructing a cause and effect diagram is an excellent way to elicit information in a very focused and directed manner. Unlike many tools used thus far, this one doesn't handle conflicting opinion and a large number of ideas will limit the effectiveness. The cause and effect diagram should be used when there is some focus to the problem or objective to be considered and appropriate experts are identified. When expert opinion is handled correctly, it is invaluable for constructing a useful diagram as well as limiting the mistakes that consume large amounts of time. The key purposes associated with the use of these diagrams are to focus attention on the most important causal activities and allow the team to assign priorities.

7.3.2 Describe the processes with matrices

Description, in particular clear description that can be easily communicated, is one of the fundamental components of planning and improvement. Graphs, networks, flowcharts and other 'pictures' from a process seem natural to use in description; matrices sometimes do not. This is probably a reflection of a mind-set that attempts to quantify things and attribute meaning to the quantities. The idea that a matrix can be used in a qualitative way at first seems novel, but becomes more natural

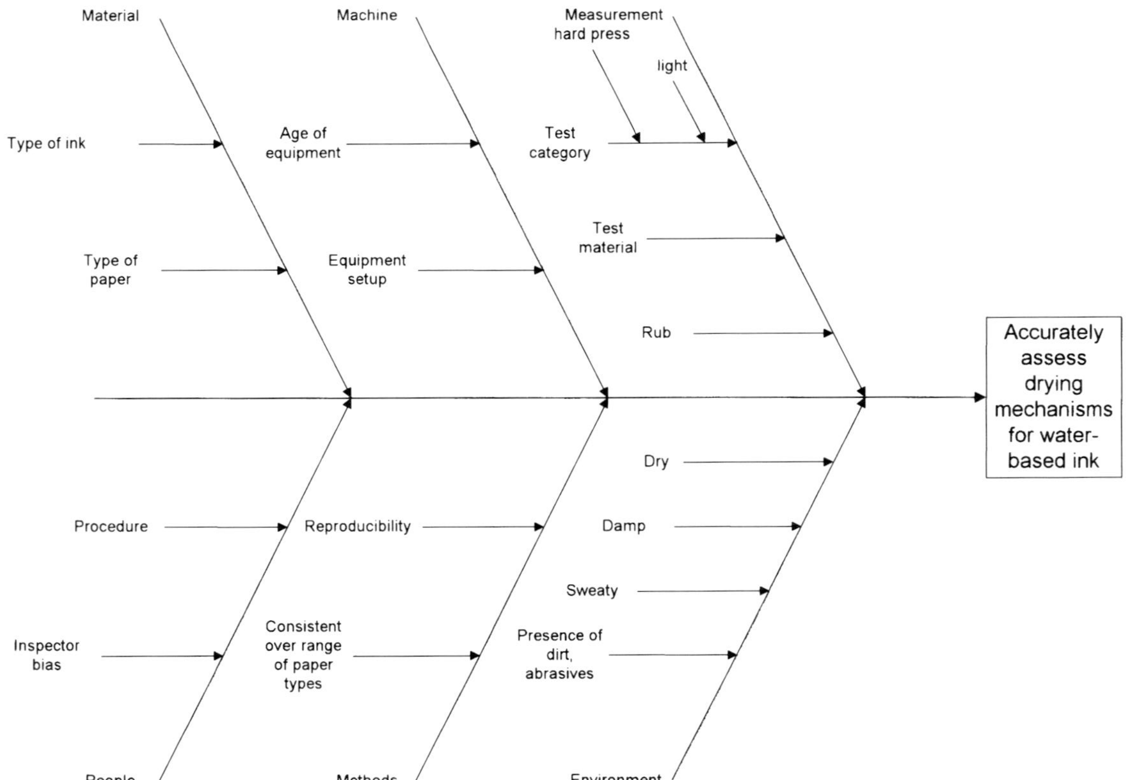

Figure 7.4 Cause-and-effect analysis of drying mechanisms

as we use it. Its use has a further, hidden benefit. Too often, people look at numbers and believe they are exact, when they are only as accurate as the measurement system allows. Replacing numbers with symbols that represent magnitude avoids reliance on a (probably false) sense of security in explicit symbols.

7.4 MANUFACTURING EXAMPLE

The quality improvement team actually used several of the tools in this phase. The fishbone diagram was most frequently used both because existing systems could have an impact on the success of new ones and because there are regulatory issues that must be adequately addressed before some of the candidate resolutions could be implemented. For example, should solvent-based ink be used, there are flammability issues that must be resolved as well as concerns for humans who will both touch the chemicals and smell the fumes. Another example centers on establishing some of the testing and comparison procedures required to assess adequately competing solutions for specific aspects of production.

Figure 7.4 illustrates the team's non-traditional use of a cause and effect diagram.

Specifically, the team was concerned with the ability to evaluate proposed techniques for drying water-based ink accurately and fairly. It is common knowledge that choice of ink has been a corporate dilemma for years. Solvent-based ink dries better but has all of the problems indicated earlier while water based ink has few of these health and safety issues but doesn't seem to dry as well. Hence, every one of the team unanimously agreed that adequate testing of alternate drying techniques was essential to success of the project. In fact, some team members clearly remembered the fiasco two decades back where a salesman from a large ink company had convinced the general manager to make a wide-scale conversion to their company's water-based ink and provided a written guarantee that the ink would work. Well, the conversion was completed and the ink used for about a month before the flood of complaints started. The ink supplier contested the 'guarantee' in court charging that F.A. Adams' Paper Airplane Division was incorrectly using the ink and it wasn't their fault. While the lawyers were getting rich, sales plummeted and the company has never quite recovered the lost sales. Hence, developing a reliable testing technique was viewed as money and time well spent and the underlying reason the cause and effect diagram was used. The 'effect' here is a repeatable and accurate assessment of the effectiveness of various drying techniques. The entire team decided to participate in this exercise because the technical experts felt that sales personnel could better define elements of the environment where the planes are really used than they could; and the sales representative knew that technical expertise was critical to development of a good scheme.

From a personal standpoint, Pi was quite pleased with the most recent turn of events. Getting the project into a form that looked a bit more traditional had improved the moral of the skeptics and this new-found appreciation between engineering and sales was almost too good to be true. Rather than make this an issue, she decided to keep a low profile and just enjoy the harmony of the group. As the diagram evolved, issues that have an impact on this test began to create an interesting skeleton to the fish. They reflected a variety of concerns, but the combination seemed to define clearly the important elements of the test and, in fact, were added to the flow diagram because development of the test program was deemed so critical and some of the issues required planning and time to execute. Examination of Figure 7.4 will provide insight into their ideas.

7.5 SERVICE EXAMPLE

One of the issues raised in the needs assessment was network connection problems. Some areas that are presently using the network have found that their network connections left something to be desired. This issue was expressed often enough that the team decided to give it some concentrated attention. One of the tools used is the cause and effect diagram.

From information in the needs assessment and from the team members' experience, the diagram in Figure 7.5 has been constructed. This cause and effect diagram is typical of those used to address a deficiency. Its 'backbone' is the effect that we want to mitigate, 'Poor Workstation Connectivity'. The major ribs chosen in this case relate to the user, the network, network software and workstation hardware. As with the flow diagram, this diagram is a tool to portray and describe the situation, to be used in further analysis as a guide. The cause and effect diagram

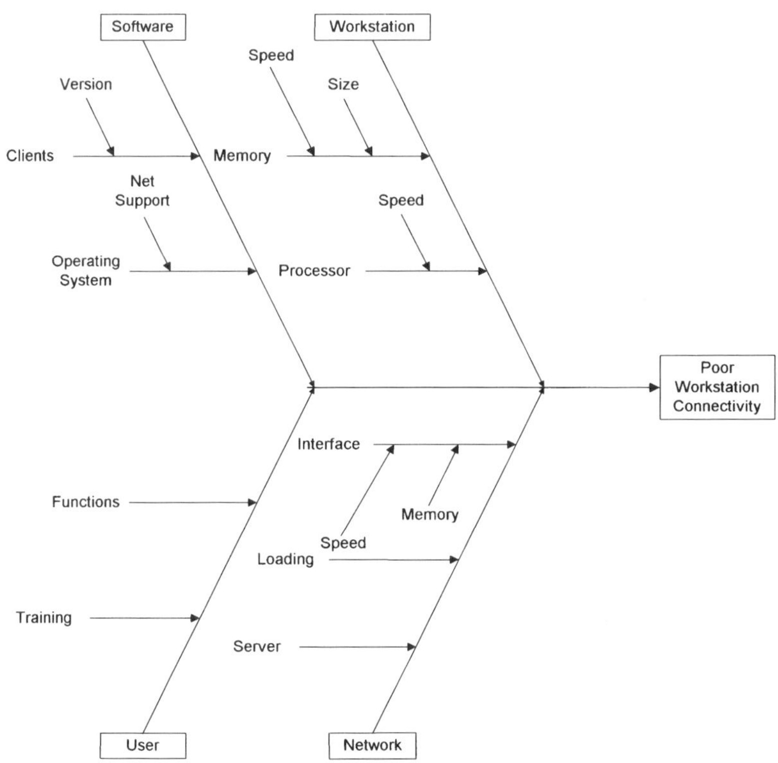

Figure 7.5 Cause-and-effect analysis of poor workstation connectivity

Allocation of Support Functions to User Type		SUPPORT				
		CONN	CONFIG	CODING	STORAGE	BACKUP
USER TYPE	APPLICATION			⊙	○	
	DESKTOP	⊙	⊙	△	⊙	⊙
	TRAVELING	○	△		△	△
	REMOTE	△			○	

Figure 7.6 Matrix: allocation of support functions to user type

does not, in itself, fix anything; it guides us in how we address the problem.

Another issue raised from the needs assessment is the team's conclusion that there is more variety among the users than was anticipated. When the team adds anticipated users, the problem becomes more severe. It appears that developing computing support for a single generic user could lead to difficulties. The team needs a way to map the different modes of use into the support services to be provided.

Whenever relationships among two or three categories of things need to be presented, a matrix diagram is a good choice. The matrix in this case shows the support categories for each major mode of use. The traveling and remote users do not presently exist at PAD, in the sense that they have computing support, but the needs assessment indicated that these modes of use are expected to develop. As the marketing and sales force evolves, in particular, the team expects these needs to become more important. Since their computing needs are expected to be quite different, they need to be identified so they can become part of the team's plans.

The matrix in Figure 7.6 is typical of matrix diagrams, in that it shows the qualitative nature of the relationships. It is easy to get bogged down in detail in attempting to completely define these relationships. The matrix diagram, with its qualitative symbols, prevents this by simplifying the level of relationship. As with other tools, this one does not provide a definitive solution; rather, it provides a guide for further activity.

7.6 SUMMARY OF KEY POINTS

- In building system knowledge deal with fact, not opinion.
- Use a tool appropriate to the knowledge, rather than forcing use of a particular tool.
- In deciding what knowledge to document, remember the 80–20 rule.

PART FOUR

Specification of Tasks

INTRODUCTION

At this stage in the planning process we begin to describe components of the system in detail. Planning to this point has been primarily concerned with describing the system, to allow more detailed work to take place within an integrated framework. Planning activities are expanding into new areas at the component level, where work actually takes place. It is important that complete and broadly shared understanding of key tasks be developed.

The key activities are development of operational definitions and specification of shared tasks. Operational definitions will document the details of tasks in ways that everyone involved can understand. While some tasks will be performed within a functional group, or even by a single individual, some will involve other functions or shared responsibility. It is especially important that these shared tasks be carefully defined and well understood. The deployment of the tasks identifies each functional or individual responsibility in the activity.

CHAPTER EIGHT

Operational definitions

In the opinion of many people in industry, there is nothing more important for transaction of business than use of operational definitions . . . An operational definition puts meaning into a concept.

(W. Edwards Deming, *Out of the Crisis*, 1986)

8.1 PLANNING OBJECTIVE: SELECTIVE PROCESS SPECIFICATION

One of the greatest causes of things going wrong in general is confusion arising from misunderstanding. We experience this in our daily lives, as well as in business and industrial organizations. Simply saying you like your eggs fried leaves a lot of room for disappointment. The same is true in instructing an inspector to reject a blemish that is too big or an operator to grind a surface until it is smooth.

The same problems can occur with project planning. In fact, unless we do something to prevent them, the misunderstandings that occur will likely be more frequent and more severe, because in planning we are more likely to be involved in non-routine activities, and tasks that are new to many of the people involved. In many cases we do not have the base of experience that frequently substitutes for clarity of specification.

Achieving clarity of specification runs into the same problem that plagues almost everything we do. That is, we would like to have clarity in everything, but we don't have enough resources. We seldom have enough time, people and knowledge to ensure that every single task and activity is completely clear. Therefore, as with other kinds of problems, we prioritize. This is the selective aspect, in that we must select those activities and tasks that are most important for clarity.

8.2 TOOLS: PARETO AND MATRIX DIAGRAMS, COMMON SENSE

These are old tools with new uses, and they illustrate both a strength of this approach to planning and a potential weakness. Engaging in this planning process involves the use of tools and methods, many of which find application both in several different planning activities and in other kinds of activities. A Pareto diagram is a method for prioritizing. It should be considered any time we are trying to select a few items from a long list. Matrix diagrams show relationships. They should be used whenever we have several items that may be related, and we wish to display those relationships. As this kind of planning process moves ahead, the people involved learn new tools and methods that can be applied in many more ways than are described here.

The potential disadvantage lies in applying this TQM approach to planning as if it were a rigid template. This is not the only place to use Pareto and matrix diagrams; it is just a useful one. Trying to keep to a rigid framework ignores two important facts: no framework can anticipate everything and every project should evolve based on the people executing it.

Common sense is not strictly a TQM tool, but it is certainly useful. In many ways, what we are about is determining, organizing and implementing common sense. It exists in the people involved in the project, and if we can find it, refine it, and communicate it well, the project will proceed well. What we need are ways to facilitate the use of common sense.

We begin by prioritizing using the Pareto chart. In the previous chapter, matrix diagrams were used to display relationships between causes and activities. These qualitative

	Y1	Y2	Y3	Score
X1	O	Δ		4
X2		O	⊙	12
X3	Δ	Δ		2
X4	Δ			1
X5	⊙		⊙	18
X6	⊙	Δ		12
X7		Δ	O	4
Score	23	7	21	

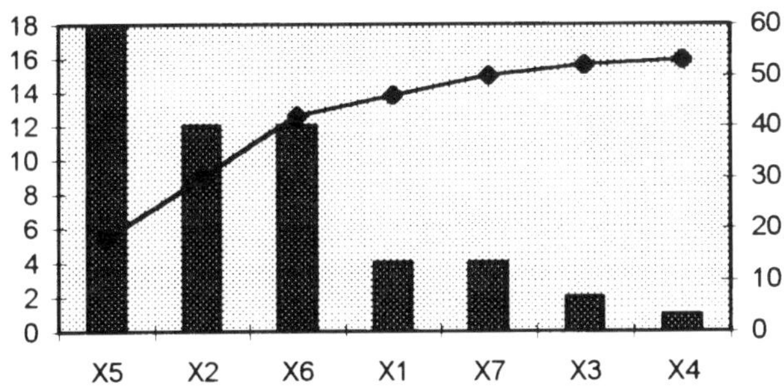

Figure 8.1 Pareto diagram in operational definition

relationships can be made somewhat quantitative by assigning scores to the symbols. One conventional scheme assigns a score of 9 to strong relationships, 3 to moderate and 1 to weak ones. With this convention (or any other that makes sense and is consistently applied) we can total along rows or columns to obtain scores for matrix factors. These scores can then be put into Pareto format as in Figure 8.1.

The figure helps us decide which activities or tasks are most important, in some sense, for development of operational definitions. The definitions themselves will take a variety of forms and it is impossible to prescribe a single format or definition of a definition that can be universally applied. What we can do is adopt an approach to development of operational definitions. One useful guide is the 5W1H model. The first four Ws are who, what, when, and where, and the H is how. The fifth W stands for 'Ask why five times', and is a common means of getting to an appropriate level of detail. This is the part of the process that is most like organized common sense, with 5W1H

	Who	What	When	Where	5W?	How
Task A						
Task B						
Task C						

Figure 8.2 Operational definition (5W1H) matrix

rooted both in the Japanese quality movement and Western folklore. An example of the use of 5W1H in matrix form is found in Figure 8.2.

8.3 SUCCESS FACTORS: CLARITY

Clarity in definition and specification is essential, and guidance to achieving this clarity is a proactive step in avoiding confusion. With several people, with different backgrounds and points of view, working on a difficult problem, the natural tendency is towards confusion. Much of the planning effort in any endeavor is aimed at avoiding chaos by directing activity. Successful direction of activity requires that every aspect of the activity be understood consistently by the people involved. Given the importance of operational definitions and the tendency toward confusion when several people are involved, this part of the planning process is certainly worth extra effort and time in ensuring that sufficient clarity exists before proceeding.

8.4 MANUFACTURING EXAMPLE

One of the concerns that had been expressed by dissatisfied customers was frequently described as 'the planes just didn't look right'. This type of feedback comes directly from the retailers who ask customers several questions before exchanging the planes, one of which is 'What is wrong with this item?' Hence, the language tends to be that of the customer who is often a disappointed child or disgusted customer. Now, FAA has always grouped the myriad of features into a general measure called 'appearance' to characterize this quality

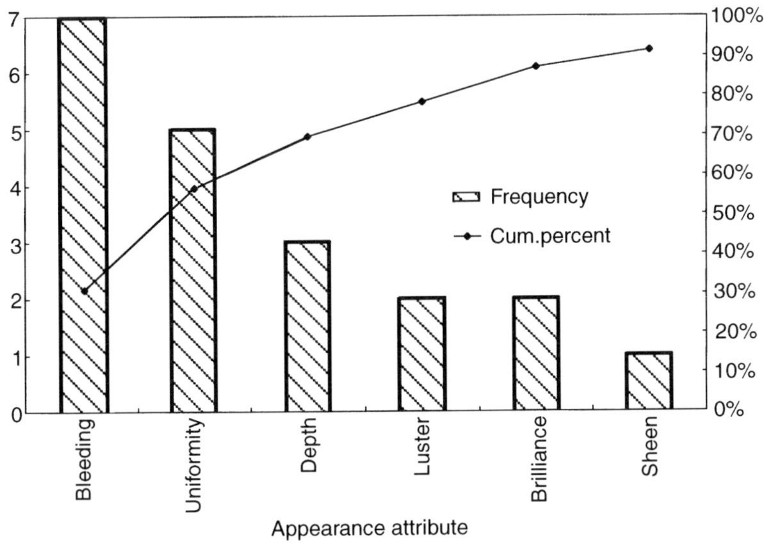

Figure 8.3 Pareto analysis of appearance

attribute. When Pi asked the team about appearance, there was dead silence followed by a variety of definitions. This discussion quickly turned into serious disagreement as each side tried to convince the other why their definition was correct.

Clearly, there is a problem here with some unfortunate implications. The customer service department had been sending customer complaint reports to the manufacturing and sales departments for years. This report identifies the number of return sales each month and, further, categorizes the complaints into one of a number of categories, including 'appearance'. Pi now understood the depth of this problem. During months where returns due to appearance demanded corrective action, everybody in the plant was fixing things to achieve a different result. It actually took the team a little longer, but when the 'dye-man' and 'operator' had the following conversation, the issue was obvious.

Dye-man: You're an idiot because everyone knows we can fix appearance by adding more solvent to improve luster.

Operator: Add solvent! It's bleeding that drives customers wild and when you add that ***** solvent, I have to work my butt off to keep that runny mess from bleeding all over the place. You haven't been adding solvent on purpose, have you? Please tell me you haven't!

Pi suggests that they contract for a customer preference survey to be conducted using comparative analysis testing. The results are presented in a Pareto chart that is shown in Figure 8.3. The team decides to take this result and develop a definition of 'appearance' not just for this project but for the company. An operational definition is duly constructed and, although happy with this achievement, Pi wonders how many similar problems are lurking around the plant.

8.5 SERVICE EXAMPLE

A problem that all computer systems have is maintenance. Maintenance at PAD is complicated because of the number of users, in particular the number of people who use either desktop or portable computers; some people use both. The systems staff agree that some maintenance should be performed by the users; indeed the more performed by users the better, since they will tend to take ownership of their systems, and the system staff will not have to dedicate manpower to maintenance.

Maintenance performed by users is not under the control of computer professionals, so some means of ensuring quality of maintenance is necessary. A team of systems staff and representative users has been convened to address this issue. Discussion quickly leads to consensus on the following points:

1. Users will be more apt to perform maintenance well if given good operational instructions.
2. Users are more apt to participate if they understand what they are doing.
3. Providing aids to the users will facilitate their maintenance.

One working aid developed by the group is a cleaning kit that can be made up by the systems staff and provided to users. By providing the kit the systems staff will ensure that users have the correct materials and will have all materials necessary. A second working aid is not an aid at all, but an automatic backup system which will perform weekly backups of the users' desktop computers on a scheduled basis, relieving them of the need to manage their own backup program.

The third working aid is a matrix that can be provided to all personnel using computers, seen in Figure 8.4.

This matrix answers the typical 5W1H questions for the two primary types of users, desktop and traveling. It provides the activities, schedule, location, reason and means for user maintenance activities. The matrix, in conjunction with the cleaning kit and necessary software, will provide users with what they need to perform the appropriate user-level maintenance. Note that the fifth 'W' is 'why', replacing 'ask why five times'; saying why is more appropriate than asking why in this case. It further illustrates the principle that these tools are to be customized in their use, not blindly adopted as is.

	Maintenance Performed by User Type					
	Desktop			Traveling		
WHAT	cleaning	disk	backup	cleaning	disk	backup
WHO	user	user	system	user	user	user
WHEN	weekly	monthly	daily	weekly	weekly	end of trip
WHERE	desk	desk	server	remote	remote	desk
WHY	extend life	data integrity	data preservation	extend life	data integrity	data preservation
HOW	user kit	user program	automatic	user kit	user program	user program

Figure 8.4 Operational definition matrix of computer maintenance

8.6 SUMMARY OF KEY POINTS

- Establish clear definitions and specifications in planning to avoid the otherwise inevitable confusion.
- The quality and scope of operational definitions are the keys to clarity of communications.
- Tools typically used for other purposes, such as Pareto charts and matrix diagrams, combined with common sense can be used very effectively to construct and communicate operational definitions.

CHAPTER NINE

Functional deployment

The key to solving the dilemma posed by tree structured organizations is to learn how to distinguish the production system from the organizational chart.

(Myron Tribus, *Deployment Flow Charting*, 1989)

9.1 PLANNING OBJECTIVE: ASSIGNED FUNCTIONS

In extremely simple cases, a single person or function has complete responsibility for an activity's outcome. This simplifies execution of a plan into a problem of daily personal management. In most cases, however, responsibility for an activity is shared across people or functions, or both. Part of specification of tasks and activities then needs to be clarity in their assignment and flow. We need to be especially careful in these cases in defining cross-boundary activities and clarifying cross-boundary processes.

9.2 TOOLS: DEPLOYMENT FLOWCHART

A deployment flowchart is similar to an ordinary process flowchart, in that it shows the interaction among process steps. Where it differs, however, is its primary value. A deployment flowchart has an additional dimension which represents the functions or people involved in the task or activity. The symbols

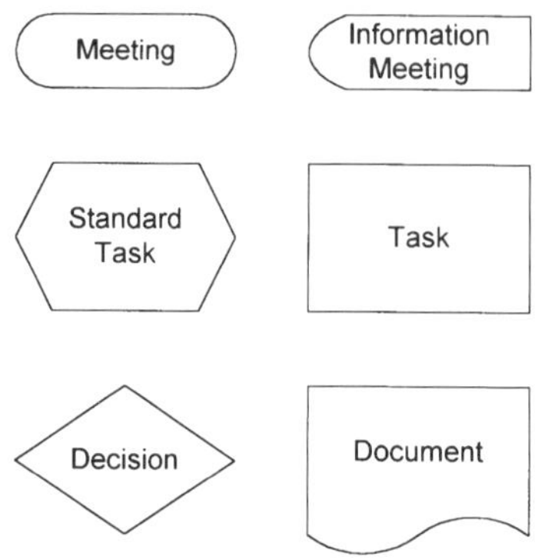

Figure 9.1 Deployment flowchart symbols

in the flowchart may be conventional flowchart symbols, or may be selected specifically to display group functions.

One useful set of symbols, used in the version of deployment flowcharting developed by Myron Tribus, is shown in Figure 9.1. The vertical line connecting the symbols at the entering point is aligned with the function or person responsible. When a task involves several people, it is stretched across the chart to span the involvement, with dashed or dotted lines under the people who are not involved. Dotted lines connecting symbols represent rework or unfavorable outcomes. Horizontal lines denote a customer–supplier relationship.

An example of a deployment flowchart is shown in Figure 9.2. This simple example describes a process for establishing specifications for a supplied material. The process begins with an engineer desiring a material with particular properties. The engineer and the purchaser first determine whether this is a stock material which is currently being supplied or readily available. If it is not, they have a meeting with one or more key suppliers to determine availability and reasonableness of the specification. If the material is not available, the engineer may need to revise the design. If it is available, cost and timing of supply need to be determined. The result should be complete agreement among the engineer, purchaser and supplier about the material sourcing. While this example is highly simplified, it

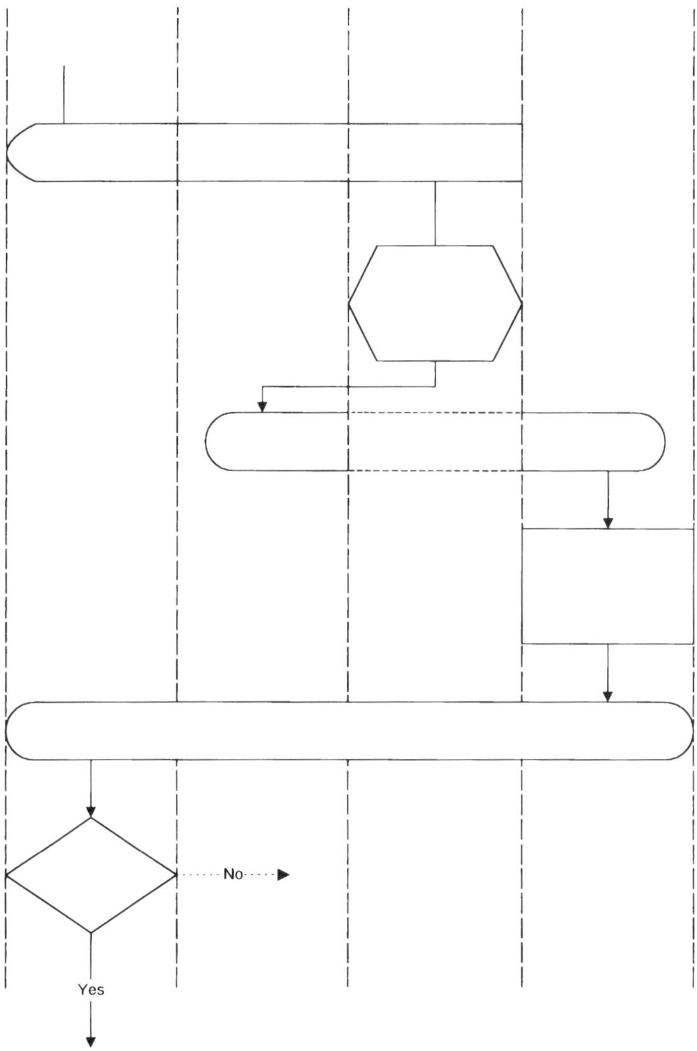

Figure 9.2 Deployment flowchart structure

is easy to see that the absence of a defined process for establishing material specifications for a new product could lead to serious problems.

Segments of flowcharts may become standardized and

replaced by a standard task or reference to a more detailed chart, leading to a hierarchy of charts. It is important that deployment flowcharts not become hopelessly confusing due to complexity and detail. A good goal is to contain each flowchart on a single sheet of paper. If it appears too large to fit, develop standard processes and sub-processes such that each can be represented with a suitable level of detail and simplicity.

Construction of a deployment flowchart is a team exercise involving all functions which would be involved in the process. Typical team problem-solving and consensus tools are used to explore the details of the process, reach agreement on its operation, and represent it pictorially. In documentation of a new process, which is frequently the case in project planning, construction of deployment flowcharts is an excellent opportunity to bring together the key players in defining the process.

9.3 SUCCESS FACTORS: PARTICIPATION OF PLAYERS

While a deployment flowchart is an excellent process documentation tool, it is worthless if it is not used and followed. Acceptance of the process is best obtained by having key players participate in its definition. In a sense, everyone is in a customer–supplier relationship with others. The development of a deployment flowchart by the team which will use the process being documented provides opportunities to explore these relationships. Just as each supplier needs to be aware of customer needs, each customer needs to be aware of the needs of his or her suppliers.

We usually think of customer–supplier relationships from the customer point of view. There is real value in considering the supplier point of view as well; that is, the supplier is really a customer in terms of the specification of what is to be supplied. Unreasonable or difficult specifications can strain a relationship just as badly as inadequate performance of the supplier. The joint development of deployment flowcharts by customers and suppliers (defined sometimes in the sense of 'next operation as customer') is a useful means of ensuring that these relationships are good ones, and that the process that is being developed will function as desired.

9.4 MANUFACTURING EXAMPLE

As Pi and the manufacturing team reach the point of using deployment flowcharting to specify tasks that must be performed and assigning them both to units within the company and to outside vendors, they are almost overwhelmed by the apparent size of the task. Almost in desperation, they begin to review the results of the previous phases of the project for some insight into how this can be done. As a group, they begin to discuss the purpose of this phase and draw some interesting conclusions. They focus on an apparent conflict in the level of detail to be displayed on the chart. On one hand, it appears that a low level would be necessary for the charts to be manageable and understandable yet, on the other hand, a purpose of the chart is to assign specific task to groups – apparently requiring much higher detail. Fortunately, this is actually a paradox rather than insurmountable problem. The key lies in the last phase, development of operational definitions, since these will allow simple deployment flowcharts to be constructed using words and phrases that each group can interpret correctly for their mission. Pi and the team now understand the extreme importance of operational definitions, even if the technique to develop them was only common sense.

A tremendous amount of information and details could be translated to various units throughout the plant as long as a particular word or phrase carried a common definition. This concept becomes so clear to the team that they actually postponed the deployment flowcharting phase until more operational definitions were made. In addition, the team found during the course of creating deployment flowchart, they would have 'sidebars' to construct operational definitions.

Figure 9.3 illustrates a portion of one of the flowcharts that was developed to specify an appropriate test for assessing drying capability among the candidate techniques. Across the

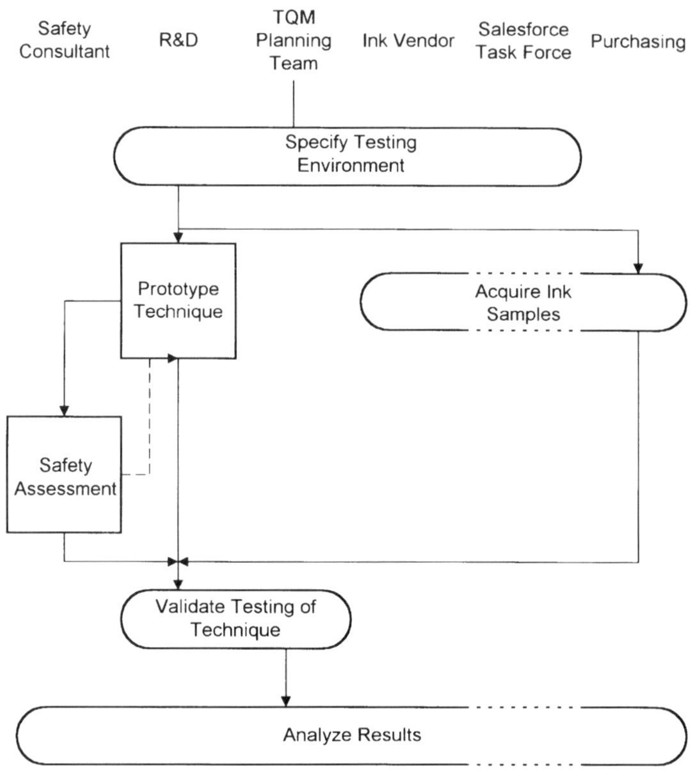

Figure 9.3 Deployment of ink testing

top of the chart are some of the units involved. Notice that these span not only internal groups but also an external safety consultant and ink vendors. A good example of the operational definition is the phrase 'specify testing environment'. Portions of the cause and effect diagram generated in Chapter 7 illustrate some of the aspects involved with actually specifying the testing environment and this flowchart indicates that these tasks must be deployed over three different groups. Apparently, then, these three words – specify testing environment – describe a multi-faceted task and contain a large number of implications. Clearly, a well-constructed operational definition is critical or there will certainly be a large number of mistakes. Among other things, for example, the operational definition indicated to the

R&D group that they must begin specifying limitations on temperature, relative humidity, etc. that must be satisfied while the same phrase tells the sales task force to specify the real environment in which the airplanes are being used. Thus, a great deal of information is conveyed in a few words and those words must be clearly defined.

The utility of operational definitions and top-down flowcharting led one of the older team members to conjecture that using this technique twenty years ago would have prevented designing automobiles that required raising the engine to change the spark plugs! 'Nay,' remarked another, 'back then it didn't take a PhD to tune a car so changing those plugs was a small price to pay for horsepower!' A newly graduated mechanical engineer was then heard to whisper to a friend, 'Tune a car? What does that mean?'

9.5 SERVICE EXAMPLE

Another activity that the team has identified from the needs assessment is upgrading of computer resources within departments. This is an activity that will involve several people in different departments, so a deployment flowchart is the logical choice to describe the process. The team has developed a generic deployment flowchart, with the involvement of personnel from affected departments, and the result is shown in Figure 9.4.

As with the matrix example, the deployment flowchart provides additional detail, beyond the conceptual data. This flowchart is intended to guide specific activities, rather than provide guidance in a conceptual sense. In this case we have users, their department heads, system consultants and the technical crew involved. While some items in the flowchart are very specific, others could be exploded into additional deployment or process flowcharts. This exercise fits the deployment flowchart well, since it involves several different people, in different departments, in shared responsibility for several activities.

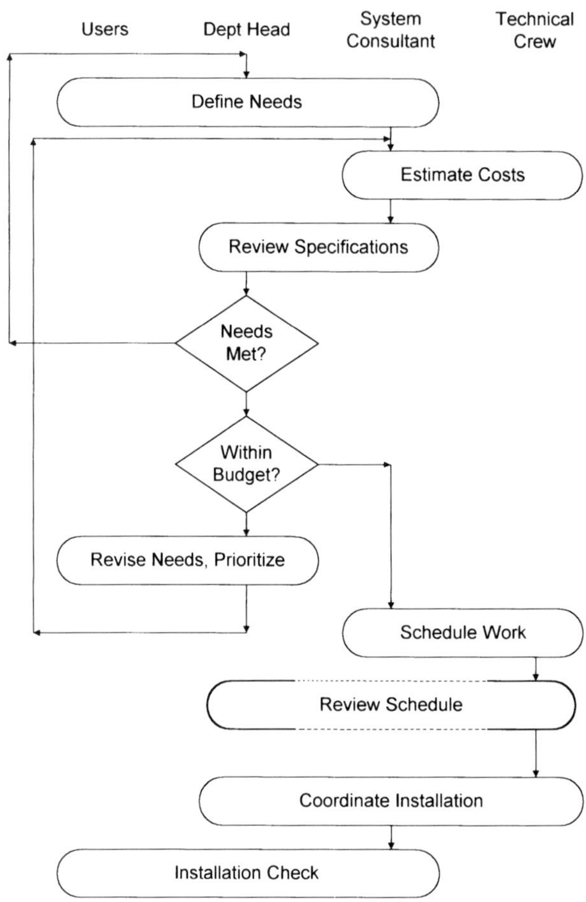

Figure 9.4 Deployment of expanding workstation connections

9.6 SUMMARY OF KEY POINTS

- Activities that cross personal or functional boundaries need to be defined in ways that all involved agree to and understand.
- The function dimension of process flow is just as important as the logic and sequence of tasks when the activity involves more than one person or function.
- In addition to their value in documentation of processes, deployment flowcharts are useful vehicles for development of customer–supplier relationships.

PART FIVE

Plan Development

INTRODUCTION

One of the deficiencies of planning in its non-TQM sense is that all of the concepts and methods to this point are preliminary to planning. Because they are regarded as preliminary, there is a tendency to focus attention on the most common physical forms of a plan, such as an arrow diagram. The result is a plan developed with inadequate preparation. With the 'preliminary' steps made a formal part of the process, they receive the attention they require, and this next step can be much more productive and useful.

The plan itself is a formal statement of what we know, understand and choose to accomplish. Previous chapters have dealt with what we know and how we understand it. The next step is to represent what we choose to accomplish in a convenient and meaningful way.

The method used to present the plan activities depends on the planning context. This context is distinguished primarily by the extent of expected risk in the future. Where the risk is minimal and activities well-defined, we use the arrow diagram. Where the risk is high or activities poorly defined, we need to take this risk and lack of definition into account explicitly.

CHAPTER TEN

Planning well-defined tasks

A seldom mentioned feature of network scheduling is its basis for a common language. Once learned, the standard terminology and symbols simplify planning communications between staff and line and among planners from different disciplines.

(James L. Riggs, *Production Systems: Planning, Analysis, and Control*, 1970)

10.1 PLANNING OBJECTIVE: DESCRIBING THE FUTURE

When planning a set of well-defined tasks, the concept is simple; the difficulty lies in the detail and magnitude of the project. Common examples of a set of well-defined tasks are assembling a collection of components to existing specifications (assembling a bicycle) or performing a number of related operations that are familiar to the personnel involved (building a house). The key test in planning a project is whether the activities are confidently expected to be familiar and routine.

In a sense, this kind of plan description is an assembly task itself, where we assemble an overall plan from familiar components. Each component is an activity or task which is familiar to the personnel involved, and we can expect each task to be completed as described. The primary variables in this case are time and physical resources. We would like to know the various times of completion of segments of the plan, and we need to link availability of resources to these times of completion.

This is an exercise in defining the future based on the past. If we have knowledge of the activities involved, through past experience or training, we can put these activities together in a form that represents their sequence and timing. A new project can be constructed from old components and defined in a way that displays the sequence and timing of activities. This is done within the generic framework of network diagrams, which include the critical path method (CPM) and PERT. Although there are a variety of formats, we will use one simple approach. A comprehensive description of these networks can be found in *Project Management with CPM, PERT and Precedence Diagraming* by Moder, Phillips and Davis.

10.2 TOOLS: ARROW NETWORK

To construct an arrow network, it is necessary to define some key terms:

node – the beginning and ending of a task which connects it to other tasks, denoted by a circle on the diagram;
task – an activity terminated by nodes and requiring time and possibly other resources to be complete, denoted by an arrow on the diagram;
immediate predecessor – a task that must be completed before a following task can begin;
immediate successor – a task that cannot start until a preceding task has finished;
source – the initial node, denoting the start of the project;
sink – the final node, denoting the end of the project;
duration – the length of time required for a task;
dummy task – a task showing only a logical connection of other tasks, included to reduce confusion in drawing the network;
earliest start – given the total time of all predecessors, the earliest time that a task can begin;
earliest finish – given the earliest start and the duration of a task, the earliest time that it can finish;

Table 10.1 Activity table for arrow diagram

Activity	*Predecessor*	*Successor*	*Dur*	*ES*	*LS*	*EF*	*LF*	*TS*
A	Source	C	3	0	0	3	3	0
B	Source	C,D	2	0	1	2	3	1
C	A,B	E	2	3	3	5	5	0
D	B	F	3	2	6	5	9	4
E	C	F	4	5	5	9	9	0
F	E,D	Sink	1	9	9	10	10	0

latest start – the latest time that a task can begin without delaying following tasks;
latest finish – the latest time that a task can finish without delaying following tasks;
critical path – the continuous path from source to sink with the longest total duration;
slack time – the extent of delay in beginning a task that can be allowed without delaying the overall project.

These terms can best be understood by constructing a small arrow network. We begin by constructing the first four columns of Table 10.1. This requires that we list all activities, their predecessors and successors, and their durations. We can now construct the arrow network from the precedence information, as shown in Figure 10.1.

The completion of the table provides valuable information that can be used to monitor the project. First, we fill in the earliest start (ES) and earliest finish (EF) times. The earliest start

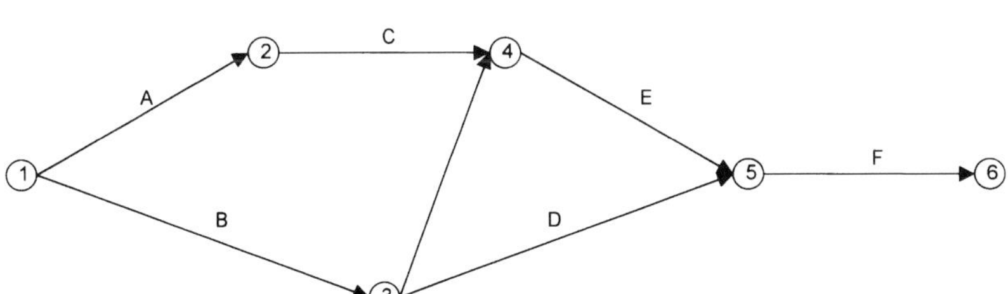

Figure 10.1 Example arrow network

is the total duration of preceding activities. Since A and B have the project source as predecessors, their ES is zero. The earliest finish is the earliest start time plus the task duration. Completing these columns is sometimes called the forward pass, since we proceed from beginning to end in the project.

The latest start time of the last task is the largest of the total durations of all paths through the network, less the duration of the last task. In this case, the longest path in time is ACEF, with a total duration of 10. Subtracting the duration of task F gives its LS of 9. Now working backward, we use the total duration of 9 for tasks D and E, since they precede F. We compute LS for task E as 9 – 4 (its duration) and for task D as 9 – 3 (its duration), and continue back until all tasks have had LS computed. The latest finish (LF) is LS plus duration for all tasks. The total slack (TS), the amount that a task can be delayed without delaying the overall project, is the difference between LF and EF.

Note that all tasks on the critical path (the longest in duration through the network) all have TS = 0. Delaying any of these will result in delaying the completion of the entire project. These become tasks of critical importance, both for monitoring their progress and for investigating whether additional resources can reduce their durations, speeding project completion.

10.3 SUCCESS FACTORS: CONFIDENCE IN DATA

The key factor for success in using the arrow diagram is confidence in the data used to construct it. For planning in this context of well-defined activities, the preliminary development of data has led up to capturing the sequence and timing of activities in this network, and its quality depends on the quality of the work in the prior steps.

Success also depends on the planning context satisfying the assumptions of the arrow network. We need to be certain that completion of all activities is equivalent to completion of the entire project, and that each activity is expected to run to completion. The variability that exists in the arrow network is variability in durations of activities and thus times of completion.

Commonly used project network software provides for identification of late activities, and impact of activity lateness and resource availability on project completion.

Finally, the size of the project is a factor. In a large project, the data management burden can be quite high. While a small project can be feasibly managed with pencil and paper, moderate to large projects (say, more than 15 activities involving more than five different resources) can be burdensome if individual activities are monitored and resource allocations modified to affect completion time. In these cases, computer software for project management is recommended. For very large projects, a top-down flowchart may be decomposed into individual arrow networks to avoid being overwhelmed by detail.

10.4 MANUFACTURING EXAMPLE

'Well, we finally made it!' exclaimed the true blue cynic. 'After six months of that touchy-feely crap, we are finally going to do what we should have done at the beginning.' This 'south end of a north bound mule' had neither missed a single meeting nor any opportunity to deride the team members about using TQM-based Project Planning and this was the last straw. The ensuing discussion wasn't lovely or even printable, but it did clear the air of half a year's worth of repressed anger. When the smoke settled, however, the value of this methodology was crystal clear. The objectives that would have never been considered had been identified, as had the critical activities that would, previously, have been considered optional but were not high priority. While the nature of the meeting wasn't pleasant and the offending member was harshly redressed by a number of people, the conclusions were absolutely on-target.

The team had correctly realized that the PERT chart (or arrow diagram) was the natural result of a great deal of focused effort in a number of activities, not of objective of the work. This is quite different from that traditional project planning where creation of the PERT is the sole goal that, when accomplished, signals the end of planning and the beginning of 'real work'.

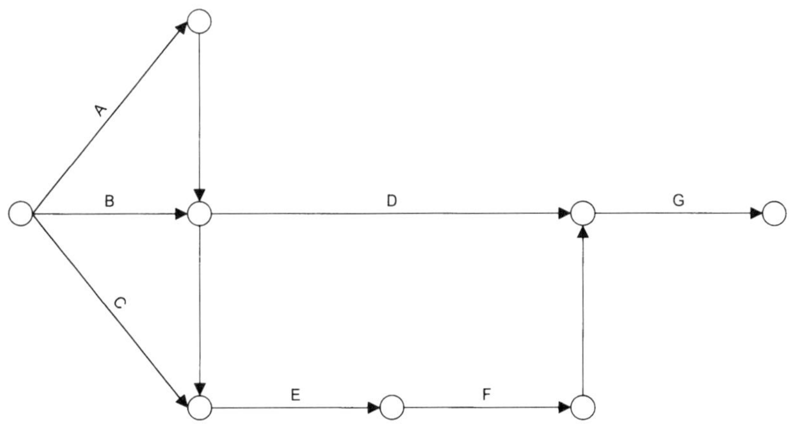

Figure 10.2 Partial network: ink selection process

One of the key points underlying the choice of this chart is confidence the team feels towards getting the individual tasks accomplished. Figure 10.2 reflects a few of the tasks associated with one of the objectives that we have been following with the team, developing a testing technique for drying ink. There is a universal confidence among the team that the tasks can be performed because team members have read of other companies who perform such tests in the trade press. Hence, a PERT diagram was selected and the portion represented in the figure illustrates that key features. Note that there are several tasks with predecessor/successor relationships. The additional information on the tasks is contained in Table 10.2.

10.5 SERVICE EXAMPLE

As the team makes progress in its work, its work becomes more detailed. The focus begins at the conceptual level and as the plans mature and narrow in focus, the tools become more detailed. Here is the case where the team has identified a generic activity that is expected to be routine, that of connecting a new department to the internal Ethernet network.

Table 10.2 Activity table for manufacturing example

Activity	*Description*	*Duration*	*ES/LS*	*EF/LF*
A	R&D feasibility	2	0/4	2/6
B	TQM team benchmark	4	0/2	4/6
C	Sales survey	6	0/0	6/6
D	Acquire ink samples	10	4/6	14/16
E	Prototype technique	4	6/7	10/11
F	Safety assessment	4	10/11	14/15
G	Validation testing	4	14/15	18/19

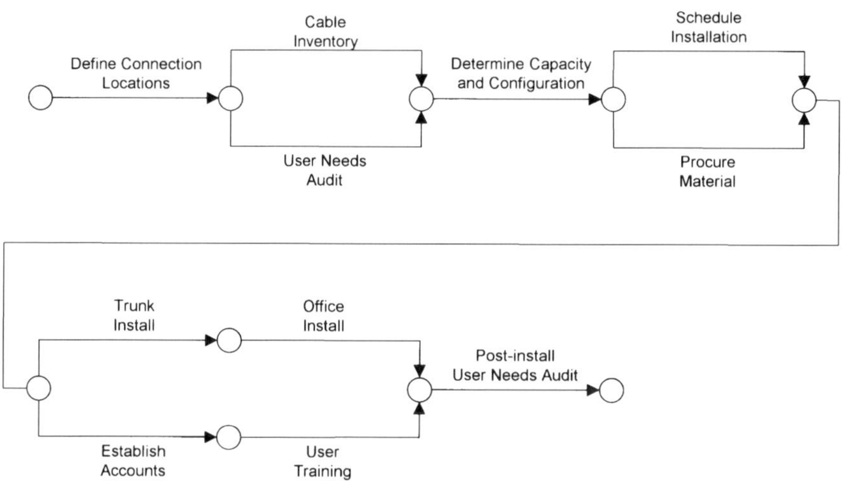

Figure 10.3 Arrow network: expanding network connections

Figure 10.3 shows the arrow diagram for this activity. This is a set of well-defined tasks. Some may take place in parallel, but all are within the scope of a single group. The team's development of this and other arrow diagrams is less a matter of arriving at consensus than determining the proper sequence of activities. By the time we have reached this level, we find the technical aspects of the project taking over. That is, we began at the conceptual level and went through those steps in order to arrive at a set of steps like this one. The conceptual planning tools help the team determine which activities should be described by arrow diagram, but the construction of the arrow diagram is guided by the technical content of the activity.

10.6 SUMMARY OF KEY POINTS

- Ensure that the arrow network fits the planning context of familiarity and certainty.
- Build the arrow network structure from a top-down flowchart and add detail based on knowledge of activities.
- Involve the personnel who will be responsible for activities in construction of segments of the network.
- Use computer aids for moderate to large projects.

CHAPTER ELEVEN

Planning open-ended tasks

What could possibly go wrong?
(Evel Knievel, describing preparations to jump the Snake River Canyon on a motorcycle)

11.1 PLANNING OBJECTIVE: CONTROLLING RISK

In many cases, in particular those involving process improvement projects or development of new functions, the assumptions of the arrow network cannot be met. In particular, we may not know with certainty that each activity in the project will be completed successfully as defined. While risk in the duration of project tasks can be accommodated in an arrow network, the conventional means of doing this still require the assumption that activities will be completed.

What is needed, then, is a method for beginning a project when we are unsure of how it will be completed. Assuming that we know the activities, proceeding with an arrow network is difficult at best and we are almost guaranteed to be changing the network structure to reflect changing knowledge. This is both disruptive to the people involved in the project and inappropriate, in that we are trying to manage risk of completion with a tool that can only manage risk of duration.

The risk we need to manage in this kind of project is whether activities will be completed successfully. Rather than assume that they will be completed, we should assume that they might not. This leads into development of contingencies, and developing plans not only to allow contingencies but to proactively choose actions based on possible outcomes and contingencies.

11.2 TOOLS: PROCESS DECISION PROGRAM CHART

The Process Decision Program Chart (PDPC) begins with the structure of a top-down flowchart. As shown in Figure 11.1, the top level is a set of macro-level activities. Since we are beginning a new process, or at least are unsure whether all of these activities will be completed as expected, we fill in the rest of the chart through a combination of brainstorming and process knowledge.

The first things to determine are what can go wrong with each activity. We wish to identify the major potential outcomes for each activity. We may break these activities down into steps in their implementation, and then identify outcomes for these steps, if this is known. The actual form of the PDPC at this stage will depend on the complexity of the project.

For each potential negative outcome, we now identify contingencies to the extent possible. These contingencies are clustered below their related outcomes. Each contingency is now evaluated. Some contingencies will usually be preferred, and these should be marked, while others should be marked as undesirable. When all contingencies have been identified and evaluated, it should be clearer how to proceed with the project, and which activities will require special attention.

This use of the PDPC will help in anticipating unwelcome events through developing contingencies in advance. The PDPC can also be used to evaluate alternatives. If there are several ways to proceed with a project, each approach could have a PDPC developed. The number and kind of negative outcomes and contingencies can assist in choosing among these alternatives. Obviously, if an alternative approach has fewer undesirable outcomes and contingencies, it should be pursued further.

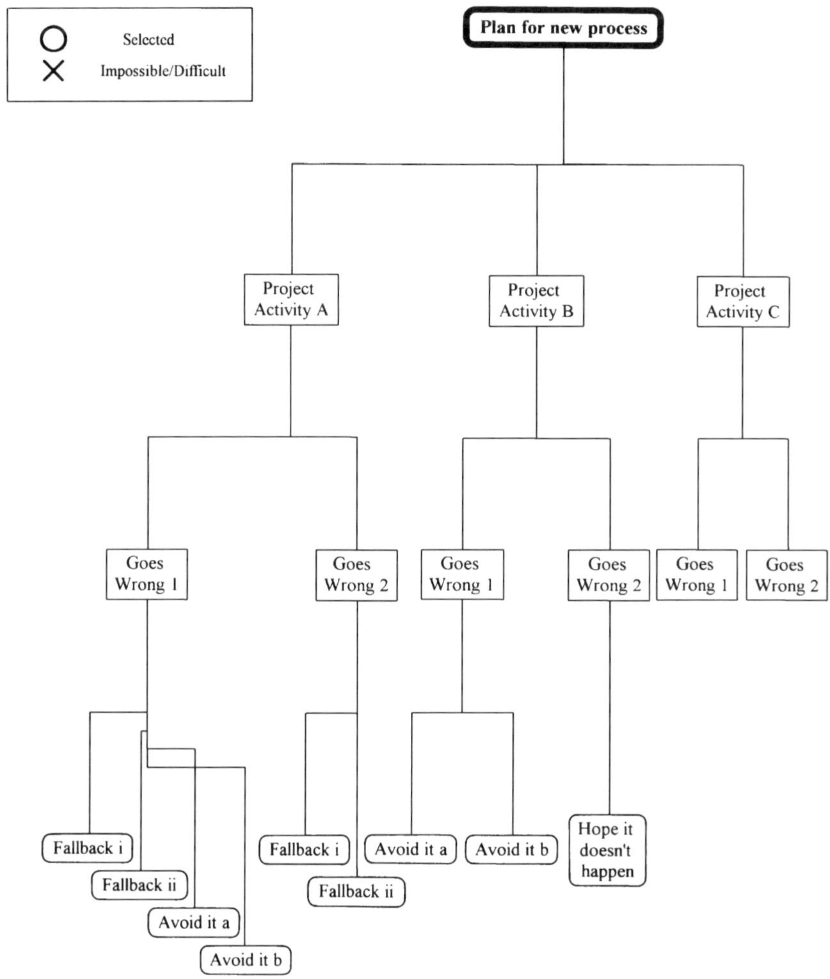

Figure 11.1 Example process decision program chart

11.3 SUCCESS FACTORS: GOOD CONTINGENCIES

While the arrow network attempts to define the future based on a well-known past, the PDPC attempts to lay out several futures for selection. The key to success is the anticipation of things gone wrong, definition of contingencies for them, and selection of both good contingencies and actions which have acceptable contingencies. The prior steps in the planning process have built knowledge about the system and its environment. We can build on this knowledge through consultation of experts, benchmarking of other processes, and, most important, involvement of the people who will be responsible for the activities.

Many of the tools used in the early stage of project planning can be used here to guide brainstorming, develop alternatives, identify gaps in knowledge and anticipate outcomes. Frequently, the knowledge and skills necessary to develop good contingencies exist; it is a matter of organizing the knowledge for reflection and bringing the people with skills together. If the project to this point has been planned with a core team, this is a good time to broaden the involvement and bring in new sets of knowledge and skill.

11.4 MANUFACTURING EXAMPLE

In addition to PERT charts used on tasks and objectives where the team has confidence that the tasks can be performed with a successful outcome, there are a number of tasks that do not fit this category. Throughout the project planning process, many team members have expressed concern about the process of applying the glue to the paper and what type of glue should be used. Unlike the ink situation or even the drying techniques, FAA Paper Airplanes has always used a standard paper glue applied using a roller and pan technique. This involves nothing more than a one-inch roller whose surface passes through glue contained in a pan on the bottom of the arc and mechanically transfers the glue to the paper that touches the roller at the top of the arc. The rpm is set by the operators based on experience – slower if excess glue is being applied and faster if insufficient glue is reaching the paper. This very basic system along with 'intuitive' controllers is quite troubling since some team members believe the process is the source of much customer dissatisfaction and the process has been regarded as 'sacred' around FAA, that is, it is not to be experimented with nor even challenged. With the high profile that this project planning team has acquired, Pi and the group decide now is the time to ask if the 'emperor really has on any clothes'.

The tasks associated with meeting this glue and glue application process objective are very uncertain and risky from the viewpoint of feeling confident that any one path will produce a successful completion. Hence, the team decides to use the PDPC as illustrated in Figure 11.2. The general idea is to outline the different paths that can happen and assess the desirability of even proceeding with each activity. In the example here, the team decided to look at all of the different types of glue and basic application techniques since each of them has a number of mostly positive outcomes. That is, the worst outcome was a realization that the glue type will not work, and this was viewed as positive since it eliminated one possibility and neither required a lot of capital nor time to evaluate.

On the other hand, Pi and the team used the PDPC on other objectives of a similar ilk and found unacceptable contingencies. At that point, these activities were either determined to be too risky to even attempt or, in some cases, it was possible to avoid the real problem with some creativity. For example, one of the contingencies involved purchase of a very expensive piece of equipment that had a very high benefit should it work, but also a very expensive price should it not because the machine was highly specialized and useful in any other operation associated with making paper airplanes. In conversations with the vendor, it was noted that a similar machine had been purchased by a company in another state. After a few telephone calls, it was determined that a unique fixture could be made by the internal machine shop at FAA for about $2,000 that would allow the contingency to be tested adequately. This modification could not withstand production speeds, but it would provide sufficient information for an intelligent decision to be made. Further, the owner of the machine agreed to allow the test for free as long as the fixture remained for use by her employees. The message that Pi carried away from this experience was that the risk of certain contingencies could be reduced dramatically using the team's creative ability and their collective 'network'.

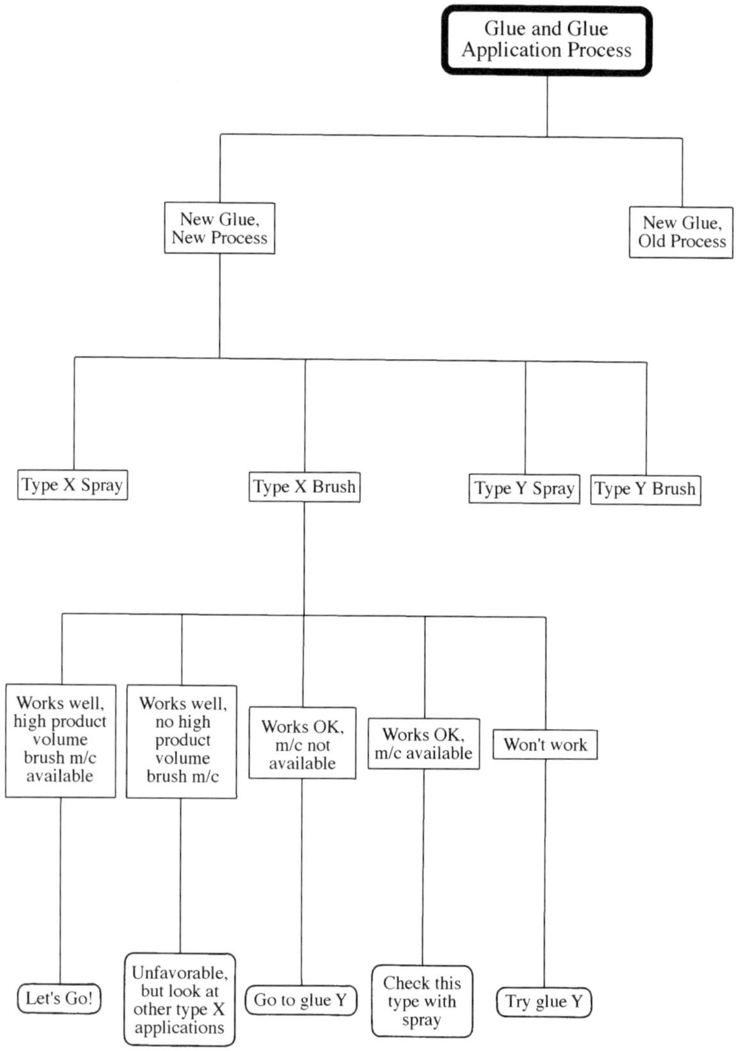

Figure 11.2 PDPC: glue application process

11.5 SERVICE EXAMPLE

Having developed a set of arrow diagrams for well-defined activities, NR and the team now find themselves facing a very

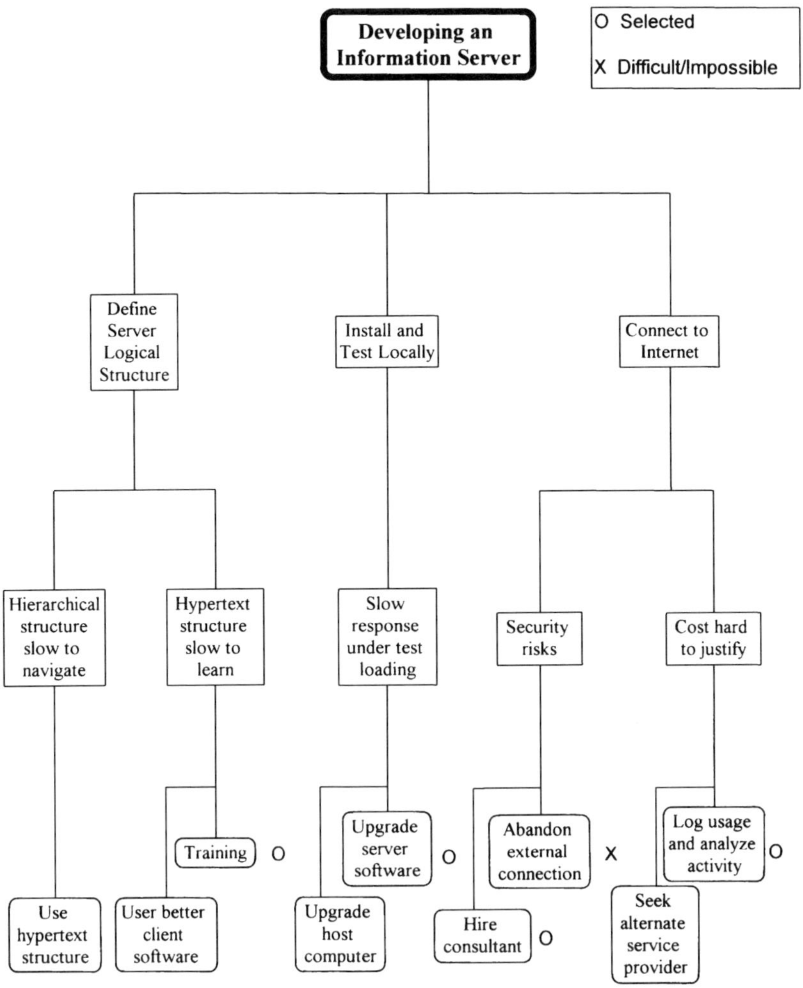

Figure 11.3 PDPC: developing an information server

different problem. Some of the preceding activities have brought attention to the need for a new information server which will be accessible both internally and externally. This would be both a marketing and information tool, serving customers, prospective customers and internal information users.

This is something new to the team, and the way to proceed is not clear. Developing this information server will consist of a set of tasks and activities, but their sequence and content is not sufficiently well known to allow use of an arrow diagram. Since there may be several alternative approaches, the team uses a process decision program chart. Their first attempt is shown in Figure 11.3.

When projects are expected to be large or complex, the team has found that it is preferred to begin with a chart that is a simpler aggregation of what is expected to be the final version. They have found it easier to develop a simple chart and then evolve it by adding detail than to attempt to develop the final chart at one time. That is clear from this example, which has only three major activities. The considerations and contingencies within the activities are also simplified. For example, the security risks with the Internet connection have alternative contingencies of hiring a consultant or canceling the project if they are severe. The team, when reconsidering the finished first attempt, will probably find other alternatives here, since one is very costly and the other leaves a set of needs unmet.

11.6 SUMMARY OF KEY POINTS

- Consider broadening involvement to include necessary skills and knowledge in the planning process.
- Use prior system knowledge to identify major activities.
- Use skill with prior tools to guide development of potential undesirable outcomes and contingencies.
- Use the PDPC dynamically as the project and knowledge about it mature and develop.

PART SIX

Implementation

INTRODUCTION

Up to this point, planning and TQM have been primarily associated with methods. That is, we have presented a set of methods that may be used in planning and described how they build knowledge that can be used in planning. Without a strong focus on implementation, however, and the dedication and structure to help it happen, the best use of methods is no more than good intentions. Implementation requires a structure, mutual understanding, action and measurement.

Another way to look at it is that there is a structure for implementation, with measurement of activities related to goals embedded within it. Organizing to accomplish these goals is the key to this structure, and the key to accomplishing these goals is their alignment vertically in the organization, such that meeting goals at one level results in equivalent goals being met at the next level. Throughout this stage of planning, it is important to keep goals realistic and well understood, and to recognize that iteration through this stage more than once may be required to achieve the mutual understanding necessary for alignment of tasks and goals to be achieved.

The implementation structure operates in parallel with the organization structure, regardless of the type of organization. For example, the organization structure may be anything from a strict hierarchy to a matrix organization to a set of teams. In all these cases, when we begin to implement the tasks in the plan developed to this point, we need a way to deploy the top-level goals of the plan and major activities into the lower-level goals of the individuals and groups performing more detailed tasks within the plan.

CHAPTER TWELVE

Implementation structure

> Thus, the means evolve from the targets. Yet many plans of action fail to include basic targets; they have only means that they then assign targets. Therefore, many are managing by setting as targets what in fact are means. This is not a good idea.
>
> (Yoji Akao, *Hoshin Kanri*, 1991)

12.1 PLANNING OBJECTIVE: ASSOCIATE TASKS WITH TARGETS

Some of the best plans fail because they are developed to the point of describing tasks to be completed, and then simply presented to the working level of the organization, leaving the implementation to the imagination of each entity responsible for a piece of the plan. This somewhat anarchic method of implementation is actually very commonly found, as it is the default mode of management in most organizations. Planners develop a set of tasks and then give them to segments of the organization, saying 'Here are your tasks. You have six months to get them done. Let me know if you need anything.' The typical result is that the various groups, teams and individuals work independently on their tasks, doing the best they can with what information they have, with little coordination. At the end of the project period, a set of tasks have been done, but the project ends up looking very different from what was expected.

In the absence of a structure that coordinates the various activities and tasks, this is the result we should expect, yet we are inevitably disappointed with it. The error here is in expecting a set of disjointed activities to become coordinated without a method to ensure that they do so. We need to associate tasks explicitly with targets relevant to the project, and facilitate achieving these targets through the ways the tasks are accomplished. Further, we need to do this consistently across all tasks and teams, giving the teams and individuals working on the tasks the proper tools to manage their own work, directed at achieving project targets.

12.2 TOOLS: TARGET/MEANS MATRIX

A target/means matrix is a self-management tool used to keep an individual's or group's work focused on the project. Targets refer to the quantifiable objectives related to the project at the next higher level of aggregation. Means refer to the tasks engaged in by the individual or group. The individual manages how the tasks are done in some measurable way and monitors how the progress of these tasks affects achieving the targets. An example of a targets/means matrix is shown in Figure 12.1.

The targets/means matrix shows the relations between targets and means using the typical relation symbols for small, moderate and strong relationships. Targets, those quantities defined at

Target/Means Matrix		**Targets**				**Check Points**		
		Target1	Target2	Target3	Target4	Number	Item	Data
Means	Means1							
	Means2							
	Means3							
	Means4							
	Means5							
Control Points	Number							
	Item							
	Data							

⊙ *Strong relationship*
○ *Moderate*
▽ *Weak*

Figure 12.1 Example target/means matrix

the next higher level, have control points associated with them. These control points may have code numbers to facilitate their accounting, and should always have a brief description and specify a method of measurement. Means, the actions that take place to result in achieving targets, have checkpoints described in the same way. A single matrix will direct the user to other sources of information, such as graphs, bar charts and check sheets, as specified on the matrix. These help in tracking progress in the control and checkpoints. An important feature of the targets/means matrix is that, while the targets are defined at the next higher level, the means are defined in terms meaningful to the user's tasks. This helps both in deployment of project goals into more detailed levels and in providing the workers with tracking methods with familiar terms.

The construction of a targets/means matrix begins with definition of targets. At the first deployment into additional levels of detail, targets will be related to the overall project. At the next level, the targets will be related to the deployed targets one level higher; they will still be aligned with the overall project, but the detail or interpretation will be more focused. This is why implementation begins with deployment of targets; the first requirement is that satisfaction of a set of targets contributes to satisfaction of targets at the next higher level, linked ultimately to overall project targets.

With targets defined, including items to measure and methods for measurement defined, means to accomplish the tasks are then defined. These are typically within the scope of the individual or group using the matrix; they are related to the specific job, task, and environment in which the work is done. The means definition also includes items to measure and methods for measurement. The next step is to define the relationships between means and targets. If a targets/means matrix shows little relationship between means and targets, the implication is that the work activities need to be redefined or prioritized. This is where the process can become iterative.

Given that the targets are rational and contribute to the project correctly, means must exist or be developed to achieve the targets. Difficulty in developing a targets/means matrix can arise from differing priorities, lack of resources, lack of training or a multitude of other causes. These indicate issues that need resolution if work on the project is to be successful. (These are also the things that will prevent success if they are not addressed. Much of the value of the targets/means matrix is identification of these barriers to success.) It is vitally important at this stage that these matrices be developed, first to ensure that the work on tasks will indeed result in the project goals being realized, and then to monitor progress as the work takes place.

12.3 SUCCESS FACTORS: LINKS TO OPERATIONAL DEFINITIONS

Lack of clarity or mutual understanding at this stage can easily arise because different sections and different levels of the organization frequently use different terms. This is natural, because their objectives and details of their work will differ. This lack of understanding is another source of project failure, which occurs when all the activities are completed, but the results differ from what was expected.

Mutual understanding is absolutely necessary, and it can be facilitated through use of operational definitions. For example, many standards refer to the use of statistically valid methods. This is not adequate, however, to ensure that statistical decisions are made consistently in different tasks. It may be necessary to specify methods, confidence levels or even statistical software packages in an operational definition of statistical validity. To take another example, it is not sufficient simply to state that each major activity be documented, leaving the methods of documentation and archiving to each section. This could result in some sections providing logbooks in file cabinets, with others providing loose notes in shopping bags.

Operational definitions need to be developed for both targets and means. While it might appear that this can add substantially to the work involved, the alternative confusion is even less attractive. Dedication and focus are required to bring this level of detail to project goals and their corresponding targets at lower levels, as well as the means used to accomplish them. This dedication has its reward, however, when the many groups and individuals in the project speak in the same terms, and valuable time is not lost in repeated interpretation or correction of activities due to lack of definition.

12.4 MANUFACTURING EXAMPLE

As the reader has no doubt realized, a transition has been made in TQM-based Project Planning from team-oriented actions to individual tasks. The first element addressed frequently begins with the operational definition and creates a set of targets that are applicable to that individual unit. From a global perspective, if all units identify their targets associated with each of the objectives and successfully complete them, all the objectives at the project level will be accomplished as envisioned by the project team. There will be no surprises. For the manufacturing example, the R&D group were especially insightful in using their target means matrix and a portion is shown in Figure 12.2.

Although this group had been represented on the team by a rather weak member, the entire group became very intrigued by the involvement projected by the team. Specifically, R&D had always felt separate and apart, 'us vs. them' as it were when dealing with the Paper Airplane division. All of a sudden, they

Prototype Testing Drying Technique	Objective Assess-ment	Cost Effective	Repro-ducible	Measure
Existing Relative Humidity Chamber		△	⊙	
Temperature Control	⊙			
Autoclave Environment	⊙	○		

Chart of reproducibility for one quality characteristic for last year

Measuring Device R&R vs Time

Figure 12.2 TM matrix: prototype testing drying technique

were on the same top-down flowchart as sales and manufacturing and, in fact, received the same task as sales and the TQM team – specify testing environment – and the output was to be a prototype testing system. Although this was clearly a unique type of task and the operational definition that accompanied the flowcharts and PERT diagrams was quite involved, the very idea that R&D would be working hand in hand with Sales towards a common goal was so foreign that the R&D people immediately wanted to do more.

Whatever the reason, their target means matrix was superb and was used as the source of this figure. As the lab manager and her direct reports began to discuss what specific targets the operational definition implied, it was clear that R&D had to be very aware of the technical issues but, at the same time, integrate other constraints on money and time. As an example, notice on the figure that there are technical issues related to testing equipment and economic impacts as well. Brainstorming led the R&D team to begin to look at several issues related to existing equipment, including the reproducibility and repeatability over the past several months and year. Further, as they constructed the diagram, it was obvious that the group was sensitive to the fact that manufacturer's claims on drying had been too frequently accepted as truthful and, possibly, was the source of some poor decisions. It became apparent from the diagram that heightened awareness of this would translate into a much better and independent testing program than before.

This implementation of target means matrix might be a bit different from that with which some readers are familiar. This is rooted in the fact that some of the units affected by the deployment might be well acquainted with these techniques and the operational definition dictates a well known set of targets. A high quality and consistent product directive might, for example, translate into manufacturing as the need for realistic tolerancing, process control, and a sampling scheme for certain raw materials. This example has purposely been chosen to be nontraditional so the reader is encouraged to use these tools in a variety of ways, including those that are different and unique. The real purpose here is to create a set of do-able tasks

	Net Use	Client/ Server Use	Spec Apps	Measure	Target
Connectivity Training	⊙			# Trained	100%
Programming Support		Δ	⊙	# Complete	# Requested
Applications Usage		⊙	Δ	# Users	# Requested
Capability Orientation		O	O	# Trained	100%

Figure 12.3 TM matrix: area computing resource coordinator

from the more abstract objective that were constructed by the project planning team.

12.5 SERVICE EXAMPLE

As the team begins to get more deeply involved in definition of activities, they encounter some that are expected to continue for some time. These may lead to new job descriptions, or new definitions of activities for personnel. The target/means matrix is an excellent way to capture these new definitions.

Figure 12.3 shows a target/means matrix for a new position, area computing resource coordinator. In very busy areas this might be a full-time position; in others these activities might be part of a person's job. (In this case, the team should develop a more complete matrix, describing all the functions done by an individual.) By convention, the left-most column contains the major job duties of the person. The relationship symbols define the importance in the mapping from the individual to his department, and the right-most columns describe the measurements and their targets. Not shown here, but existing at the next higher level, would be the matrix for the department or the employee's supervisor; this matrix would have the column headings shown here in its left-most column.

12.6 SUMMARY OF KEY POINTS

- Deploy project goals into more detailed targets for implementation.

- Give the working levels tools to manage themselves, within the context of the project.
- Adopt operational definitions for targets, means and other critical issues within the project.

CHAPTER THIRTEEN

Plan alignment

> The function of a manager is to make his workers successful, and vice versa.
>
> (origin unknown)

13.1 PLANNING OBJECTIVE: MEETING UPPER-LEVEL GOALS

The previous chapter described the structure for implementation, leading to alignment; we now turn to achieving alignment within this structure. The recognition of this alignment, from the workers at one end of the structure to the goals at the other, is one of the most powerful factors in making significant progress in any endeavor. Too often, the different functions within a project or program seem disconnected one from another, with each going its independent way. Each individual needs to see and understand the linkage between what he does and how the overall goals are met.

The deployment process begins this recognition with simple but powerful tools that provide the model. The model needs to be made real, through better understanding and through tracking of processes as they take place. In the initial stages of the alignment process, the primary need is understanding. This involves tracing the deployment of functions through the organization or team, to make the links between functions and up and down the hierarchy explicit. As these links become explicit, we achieve the realization that, through alignment of effort within a group, the group goals are met. With alignment of effort among groups, the larger group's goals are met. As we proceed through the project structure, we find an alignment of goals from the overall project goal down through the project organization, the goals becoming more explicit and focused at each succeeding level. What we seek to achieve is the alignment of these goals on an operational, continuing basis, in both directing work and measuring it.

13.2 TOOLS: FLAG SYSTEM, CASCADE DIAGRAM

The primary tools in achieving this understanding and closing the loop with measurement are the flag system and cascade diagram. The project structure can be thought of as a set of inverted funnels that interlock. Goals are deployed from the neck down, and measurement of results is aggregated and passed up the neck to the next level, where it may be aggregated again.

The cascade diagram illustrates both of these flows. Its primary use is to show the interconnections of a set of levels of a project, where each level can be modeled in a deployment matrix. While the deployment matrix provides detail for use within a level, the cascade diagram shows interconnections through several levels, providing the chain of logic from top-level goals down to operational details.

The flag system is primarily concerned with measurement of processes, and is most effective when used by an individual manager in tracking functions within his area of responsibility. It provides a graphical display of measurements on these functions, both individually and in the aggregate, overlaid on the structure that describes the relationships among the functions. It is assumed that each function will have its own measurement system and goals, and that these are maintained by the individuals involved in it. The flag system does not replace that system; rather, it provides the overall manager of several of these systems with a graphical tool that provides a snapshot of performance and recent history.

13.2.1 Building a flag system

The simplest flag system is just a sheet of paper with a line drawing of the function structure, overlaid with graphs of the most important performance measurements. The most important factor in its design is selection of the measurements to be tracked. The number should be small, reflecting the principle that only a small number of things should have a high priority, and that only the highest priority items should be managed closely.

The next level of sophistication, and the one that is likely to be most useful, is implementation of a flag system in a spreadsheet computer program. The flag system in Figure 13.1 was produced in such a system; the major requirement is the ability to present tabular, structural and graphical data easily and conveniently. Each branch of the tree has a similar measurement and graph associated with it. The kind of graph would naturally depend on the kind of measurement. The 'trunk' graph is the aggregation of the branches. The branch graphs can easily include target values, as well as plots of performance over time.

13.2.2 When to use a cascade diagram

The cascade diagram is more a tool to inform than to manage. Its first use is in the early stages of the project, when we seek to develop understanding of functional relationships and the links among functional and overall goals. It is developed as the project develops, and provides a key means of communication throughout the duration of the project. Having used cascade diagrams at the outset, it can be useful to post them in strategic locations, as visible reminders of project relationships. An example of a cascade diagram is shown in Figure 13.2.

13.3 SUCCESS FACTORS: BREADTH WITH FOCUS

As we go deeper into the implementation details of a project, we must take care to avoid over-management. The level of detail

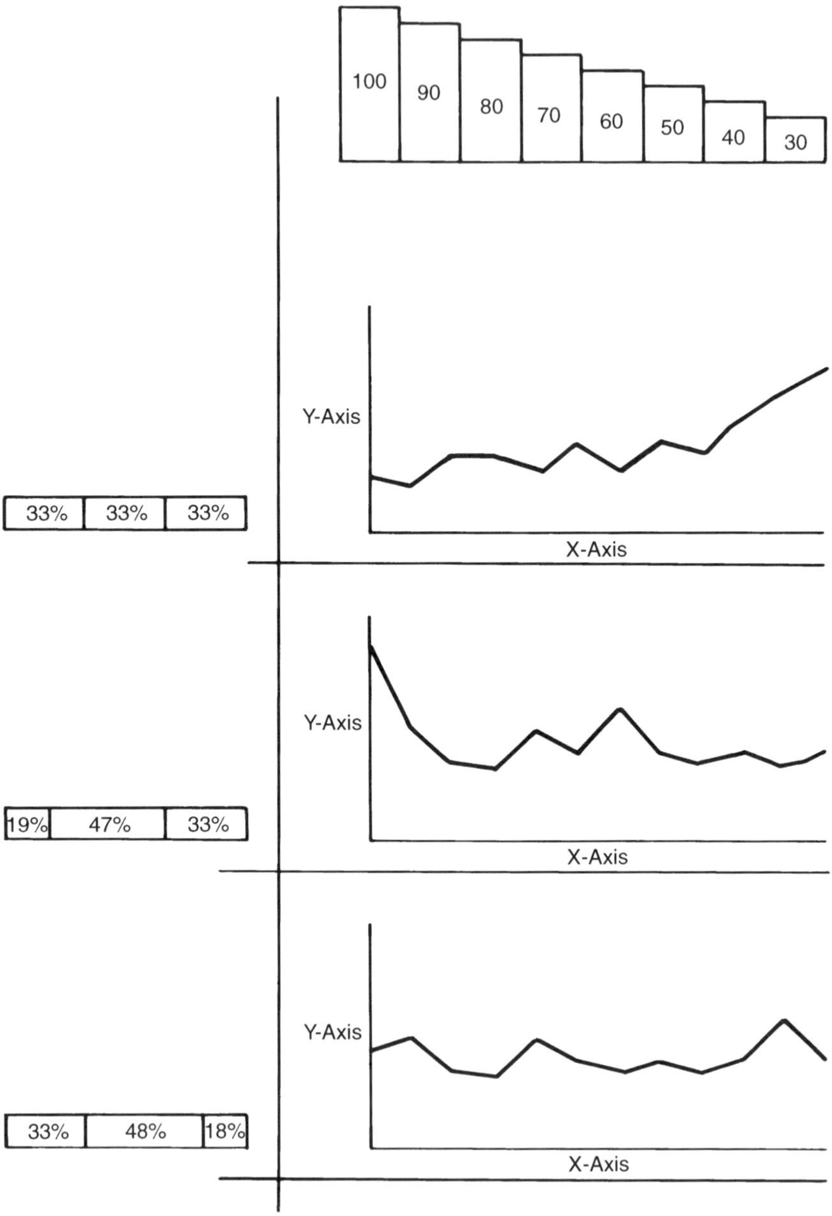

Figure 13.1 Example flag system

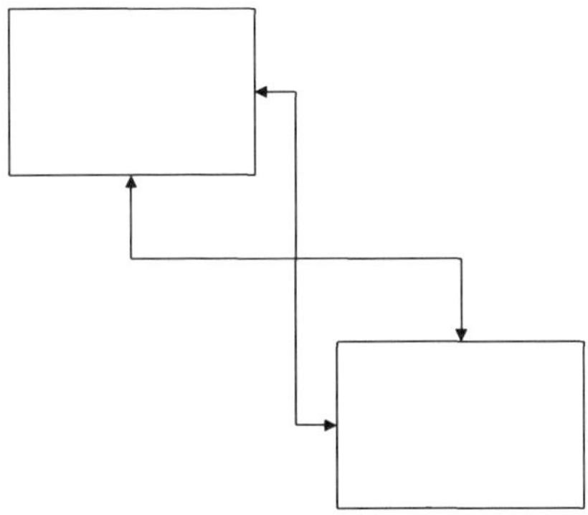

Figure 13.2 Cascade diagram structure

increases, and the ability to exercise direct control correspondingly decreases. Controlling all project functions directly becomes impossible. The principle of alignment allows focus to be maintained on critical or high priority items, while maintaining breadth of control. This can happen only if the various functions within the project are aligned, in the sense that they understand the links between their personal goals, the goals of their functions, and the overall project goals. This facilitates a mode of self-management within an overall structure of project goals, that allows us to use the exception principle productively to manage only what needs management, and ensures that what needs management is itself at a manageable level.

13.4 MANUFACTURING EXAMPLE

The flag diagram illustrated in Figure 13.3 was used by the project manager as some of the recommended changes were installed in the production facility. The Paper Airplane Division does not have a scale-up facility so any modifications to the product or process, even if they are lab tested and approved, must be carefully monitored as they are introduced in the full scale production environment. Since the first phases of the

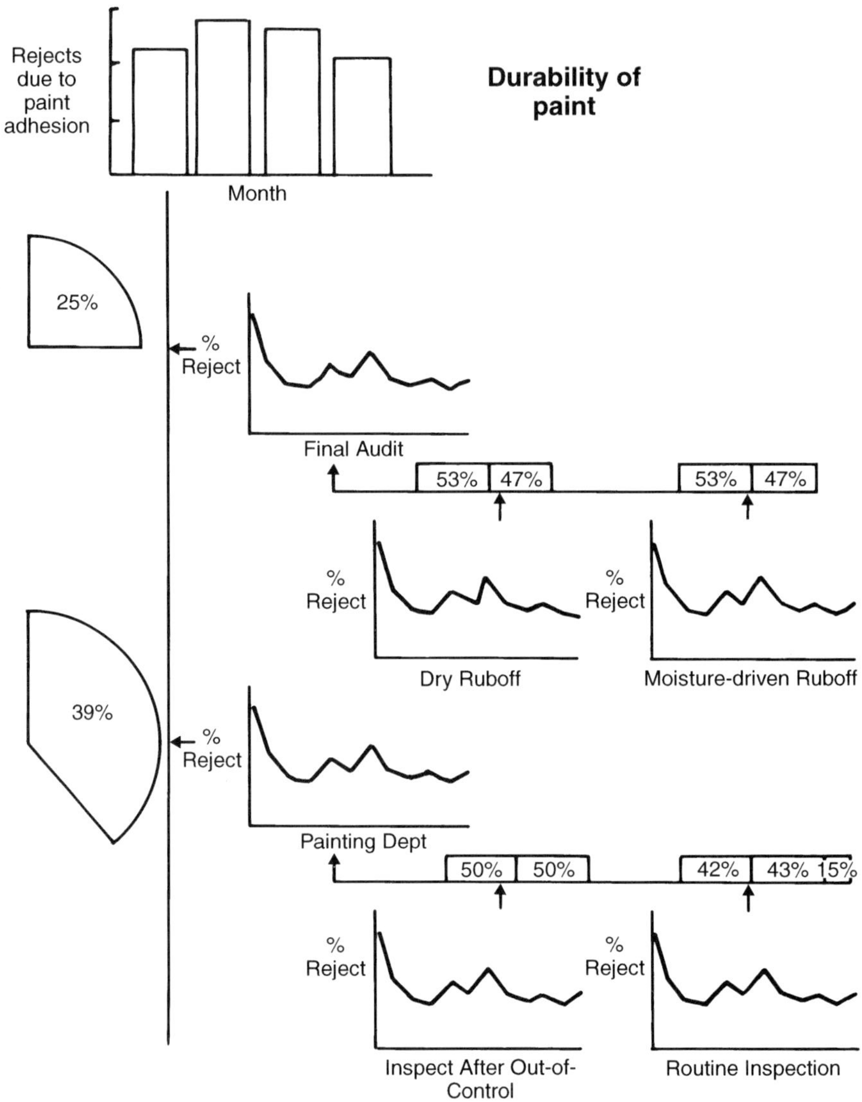

Figure 13.3 Flag system for paint durability

project had both uncovered several problems and provided a number of resolutions, the planning team and project manager were faced with measuring success, with the full knowledge that some of the changes would interact with each other and that all would probably behave differently in the full-scale production environment than they had in the controlled tests. One of the ways that they defined success was in terms of the overall reduction in the number of planes that were declared defective due to ink durability related problems. Hence, the team worked with a number of groups to immediately install a number of tests throughout the process to measure this defective percentage.

Everyone wanted to gather data with the existing system so this could be used as a baseline from which any changes could be compared. In addition, the TQM project planning team was statistically literate and, as such, convinced all the others that it was desirable to collect sufficient information for statistically valid conclusions to be drawn. The flag diagram in Figure 13.3 illustrates how and where these data were collected and two levels of aggregation. What the diagram doesn't show is the planning behind the construction of this system. In general, the planning tools described in this book were used first to define goals and, eventually, transform them into measurements. It was important, for example, that the data not only indicate the number of defectives as the old system did, but for the new system to provide direction concerning where the defectives were being generated as well. Although this is perfectly sensible, historically it had never been done because the extra inspections both consumed resources and provided the basis for punitive actions by upper management.

As silly as it sounds, upper management was accustomed to 3% defectives due to ink durability problems and didn't react provided it was not shown that a single department caused the problems. In a sense, sharing the blame maintained the status quo – and that status quo had nearly bankrupted the division. Anyway, specifying the locations and tests to be performed was another thread of this project that the TQM project planning team developed and the project manager executed. The successful implementation as indicated in the figure provided critical feedback during implementation and full-scale testing of the changes.

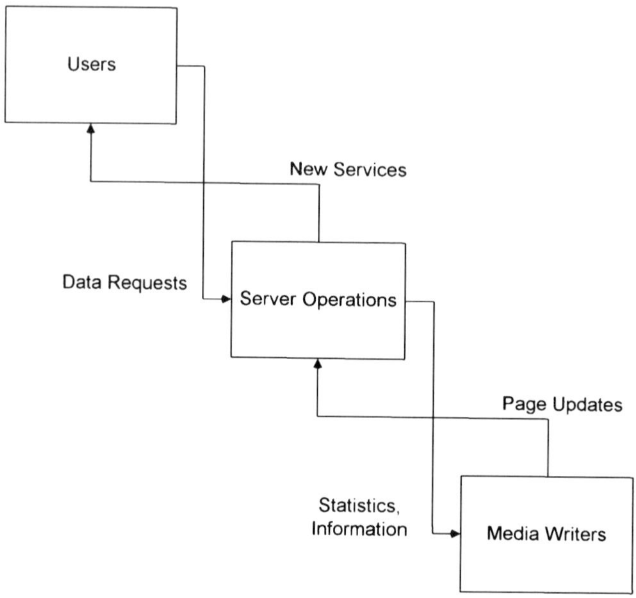

Figure 13.4 Cascade diagram for server maintenance

13.5 SERVICE EXAMPLE

The information server described with the PDPC will need maintenance once it is operating, and one aspect of its maintenance is file updating. What we have here is the interaction of three different functions. Users access the server and may request specific data. The server is operated, maintaining logs and statistics on data requests and forwarding requests for specific data. The server media authors respond to the logs, statistics and data requests, and refine the data available or develop new resources based on usage and requests.

This process is shown in Figure 13.4. In using a cascade diagram, its function must be kept in mind by the team. This diagram does not describe a single job function, nor does it provide sufficient detail to describe a process. What it does well is describe or document a set of relationships within a hierarchy of functions. In this way, it can serve as a reminder of how these functions are related.

13.6 SUMMARY OF KEY POINTS

- Relationships among goals at all levels need to be explicit.
- Develop a structure that needs minimal detailed management because many functions are amenable to self-management.
- Strive to keep the number of functions or measurements directly managed small.
- Successful functions lead to successful projects, if the organization is aligned.

PART SEVEN

Project Review

Introduction

From the Shewhart cycle to the latest version of ISO9000, continuous improvement is a vital part of any process. In the context of project planning, plan review is the way to incorporate feedback into the planning process. This brings up the use of checkpoints and control points, and mechanisms to review conformance to plan and provide corrective action.

The PDCA cycle certainly has a place here, and this is also the best place to begin assessing the effectiveness of plan deployment. As with the overall planning process, we first plan how to do it, and then we do it. As before, we use common and simple methods that do not unduly add to the project effort.

CHAPTER FOURTEEN

Planning for review

No battle plan survives contact with the enemy.

(military maxim)

14.1 PLANNING OBJECTIVE: GET READY FOR ACTION

In military science, the leaders in an action have responsibility for developing the battle plan and then, as the action takes place, accommodating the plan to the changing environment. In project planning, we should not regard the planning and project environment as the enemy; we should, however, recognize that the environment can change, and that the execution of the original plans can have effects that differ from what was anticipated. One of the key roles of the project manager is in coping with the dynamic nature of project execution.

Coping with dynamic execution begins with good preparation, and both the responsibility for and use of this preparation lie primarily with the project manager or management team. Preparing for project execution builds on the documentation developed throughout the previous steps in project planning. Critical control points are identified; these become the basis for a system of project tracking. As personnel involved in the details of project activities are monitoring their checkpoints, to track the progress of their task details, information is being gathered from control points to assess the macroscopic progress of the project.

The project managers cannot possibly track every detail of project execution. One of the purposes of developing project documentation described in previous chapters, in particular deployment charts, target/means matrices, and flag systems, is to provide the basis for project tracking. These tools help identify critical activities and their control points. These form the information basis for project tracking and review, in a simple system that facilitates project tracking with minimal effort. Other common tools are then used to process this information.

14.2 TOOLS: CHECKLIST, MATRICES

Checklists and matrices, the workhorses of continuous quality improvement, find additional usefulness in project review plans. An important principle in project review is simplicity, and capturing review plans in simple tools enforces the discipline of simplicity in project review. These tools are used here in two important areas of project management – managing resources and managing deployment.

Within the details of task management, the resource issue is one of allocation. That is, we are primarily concerned with the extent of a resource and its assignment to project activities. In preparing for project execution and review, however, we shift from a 'how much' orientation to a 'what' orientation. At the top level, our concern is more about what resources are required, to ensure that the proper type of resource is available at the various stages of the project.

A checklist in matrix form for resource planning is shown in Figure 14.1.

This checklist identifies critical resource types, and shows the functional requirement for these resources at various stages of the project. In this particular 'T-shaped' matrix, we can display both the requirement of resources over time (by stage of the project) and the requirement of resource type by project function. Resources specific to project tracking, review and management need to be shown here as well; these are some of the resources most likely to have been ignored to this point,

STAGES				FUNCTIONS		
			R			
			E			
			S			
			O			
			U			
			R			
			C			
			E			
			S			

Figure 14.1 'T' matrix as a checklist

since the emphasis has been on the development of the project activities, rather than their review.

Deployment was described in previous chapters in terms of flowcharts involving several functions and deployment of activities through the target/means matrix. Deployment in project review is a deployment of the review function, tied to the deployment of activities. This can also be captured in a matrix, as in Figure 14.2.

In this matrix we display the collection of critical control points in terms of the review activities that address them. One of the most onerous tasks in any operation is reporting, especially when reports have little apparent relation to the activity performed. Careful design of the review plan to ensure that reports address the proper issues in a form that does not add unnecessary burden is one of the most important functions of the review plan.

	Review Reports			
C P				
O O				
N I				
T N				
R T				
O S				
L				

Figure 14.2 Deployment matrix for reports and corresponding controls

14.3 SUCCESS FACTORS: IDENTIFICATION OF GAPS AND CONFLICTS

Planning for project review is more than just a means of gathering data. It also provides one more check on project integrity and success. A potential weakness of this approach to project planning, which depends heavily on deployment of responsibility for activities and their plans, is that the top-level or system-level issues can fail to be addressed properly. This is in contrast to a project management approach that has its primary focus on top-level issues. In this case, a frequent failure is to focus on top-level issues to the extent that little planning is done with regard to the activities themselves. Project management without explicit deployment can easily fail because the plan stops too soon, leaving more detailed plans up to the activity level personnel, but providing little guidance. Deployment-based project management can focus on the activity level to the extent that system considerations are neglected.

The project review plan provides an opportunity to resolve these system-level issues, which typically manifest themselves in the form of resource gaps and conflicts. Timing and sequence problems are generally resolved well through the project flowcharts and activity networks. Resource assignment, however, remains primarily a management function. The use of checklists and matrices to form the project review provides feedback from the activities to the system level, allowing one more review of the resources and their allocations.

This avoids the apparent lack of common sense found in some projects. While it might be only common sense that a resource is required for an activity, the common sense that ensures the resource is present requires some guidance. Too often, assumptions made about common sense and common knowledge are responsible for project failures because the project review plan was not adequately developed.

14.4 MANUFACTURING EXAMPLE

It is important to note that the process of project planning is nearly complete. This is the home stretch that can be viewed in much the same way as a race or an athletic contest. Specifically, a lot of work has been accomplished and things are going smoothly as you pay attention to detail; however, all can be lost down the stretch if that same perseverance is not continued until the very end. It is easy to gain some feel for the amount of information that has been processed by again reviewing the problem statement and thinking about addressing every one of those issues with the same dedication and effort using the tools outlined in this book. In short, our team has accomplished an extraordinary task, even to the point of installing measures in the process that will monitor product improvement during scale-up.

Attention is now turned towards the final tasks of seeking to ensure that the project actually is successful and the plan is truly executed as outlined. One way to encourage this is to identify the group that is responsible for ensuring that each of the tasks is accomplished and which groups should provide support. A matrix diagram is frequently used to portray this information during negotiations and Figure 14.3 illustrates a portion of the matrix used by this team. In this matrix, the key tasks and departments are identified on the sides of a standard matrix format. The symbols are redefined as follows: ⦿ = primary responsibility, ○ = high-level supporting role, △ = low-level supporting role. This matrix can actually be used by several levels within the organization because not only can the PM quickly communicate the responsible unit for each major activity but the key players in those units can easily determine who is providing their primary support.

14.5 SERVICE EXAMPLE

The team is now taking a look backward at what it has accomplished, and thinking about how to ensure that the

Responsibility Matrix	R&D	Sales	MFG-Stage 1	...
Drying Techniques Test	⊙		○	
Alternate Application Technique		○	⊙	
Customer Information	○	⊙		
...				

Figure 14.3 Matrix of divisional responsibility

Deployment review	Hardware	Software	Network	Training	Facilities
Director of Computing Support	⊙	○			△
Area Computing Resource Coordinator		△	○	⊙	
Area Director				△	△
Area User Coordinator	△	○	△	○	
Facilities Maintenance	△		○		⊙

Figure 14.4 Allocation of responsibility in computing support

project continues to succeed. One thing to be done is to define clearly responsibilities in some key areas. The matrix in Figure 14.4 shows the deployment of key functions to areas with corresponding responsibility.

This matrix shows the allocation of responsibility both within the computing support operation and in other departments which have a role in some aspect of computing support. For example, the facilities maintenance group might have a tendency to forget completely about their role in computing support, since it is not in their primary responsibility. Gaining consensus among those involved for development of this matrix helps document the shared responsibilities.

14.6 SUMMARY OF KEY POINTS

- Use simple and common tools to prepare for project review.
- Use the review plan as a final check on gaps and conflicts in resource utilization.
- Keep the review plan simple and keep reporting requirements to the minimum necessary for project tracking.

CHAPTER FIFTEEN

Review structure

> The best chieftains develop the ability to ask the right questions at the right time.
>
> (Wess Roberts, *Leadership Secrets of Attila the Hun*, 1987)

15.1 PLANNING OBJECTIVE: BEGIN EXECUTION

The beginning of execution is the beginning of the review cycle, which can be viewed as asking the right questions at the right time. The project manager at this stage, having developed a review plan based on the overall project, becomes in large part a facilitator; a fine balance is maintained between sufficient involvement in activities to support them and sufficient detachment from them to allow individual management of tasks. The primary resource of the project manager is information about the project, used both to track project progress and to support project activities.

15.2 TOOLS: PLANNING DOCUMENTS

The documentation set for a project can vary widely, depending on the scope of the project, and not all documents developed are useful in guiding the progress of the project. (Some are primarily useful only within certain activities or only to certain individuals in the project.) Some documents communicate externally to the

Table 15.1 Typical documentation set

Tree diagram – overall structure
Matrix – key resource allocation
Arrow diagram – overall project flow
PDPC – new process flow
Deployment chart – selected processes
Top-down flowchart – overall project
Target/means matrix – project evaluation
Deployment matrix – review guidance

project, and others internally. Some of each of these should be archived. The actual documentation set for project review should be defined by the project, rather than an arbitrary specification.

A typical documentation set is found in Table 15.1.

In this table we find project documentation developed using the tools described in previous chapters. These are primarily descriptive documents at the top level of the project. Detailed documents having to do with execution of activities are certainly needed within their activities, but they are not especially useful in overall project review. We also find documents in this set that are used for external communication; these may serve as the basis for management updates and communication to others outside the scope of the project, as well as to communicate overall project progress to project personnel.

15.3 SUCCESS FACTORS: EFFICIENT REVIEWS

The review plan and documentation represent one vital component of project review. These comprise what must be known before beginning the review process. The other vital component is review timing.

Timing of the review process, in particular scheduling review meetings with project personnel, must be carefully designed. A good starting point for this design is the set of control points identified in the review plan. Each control point should be accompanied by information sufficient to determine how rapidly the control value might change and thus how often it should be updated. There is no point in updating information more often than the data meaningfully change. Some control points may signal completion of activities that must precede other activities; these completion times or dates may also indicate useful times for project review updates.

The project time line itself can provide guidance in its review and updating. For example, in a project with a duration of less than a year, updates should take place at least monthly, and more often for critical activities. If updates are monthly and a control point indicates significant delay at a monthly update, two full months can easily go by before the control point's activity is brought within schedule; this could be a substantial portion of the project life. Conversely, weekly updates for a long project with slowly changing control points will generate many reports with little meaning.

Rather than have a strict schedule of meetings for all project activities, it is preferred to schedule updates selectively based on the criticality of the activities and their internal time lines. It is also useful to have in mind an action plan for dealing with slipped critical activities before they are encountered. This could range from more frequent updates to overall review of the activities themselves. Not all activities and tasks will have the same criticality and time sensitivity.

15.4 MANUFACTURING EXAMPLE

The final phase of project planning illustrated in the manufacturing example concerns addressing the dreaded paperwork portion of a project. Frequently, the current project management method demands that the project manager construct monthly reports from scraps of paper from engineers, technicians, salespeople etc. who are 'too busy working' to bother with writing reports. Anyone who has ever been a PM placed in this position knows that this task is simply dreadful and

Document Preparation	R&D	Sales	Engr	MFG-Stage 1
Customer Profile	O	⊙		
Baseline Database			⊙	⊙
Testing	⊙		O	O
Improvements-Hardware	O		⊙	
Improvements-Process			⊙	⊙

Figure 15.1 Documentation responsibility for manufacturing project

time-consuming, and the resulting report is frequently useless. The impact of this situation frequently isn't fully understood until a few years have passed and another project team begins to examine the changes and finds a final report that is worthless – no drawings or drawings that aren't accurate, insufficient descriptions, woefully lacking calculation files etc.

To avoid all of this, the TQM project planning team decides to hold a meeting of all department managers and gain a consensus on who is ultimately responsible for the key portions of the document. Figure 15.1 illustrates a portion of the matrix that was used as the mechanism to gain that consensus. The beauty of this approach was that department managers could see exactly how the responsibility was divided and self-regulate both those managers who didn't want to do anything and those who felt they should write the entire document. Interestingly enough, this meeting took less than half an hour and, afterwards, the PM visited each department manager and acquired the name of the individual responsible for that reporting. Clearly, the PM was ultimately responsible for getting the documents assembled and cohesive, but the need for information was now deployed to the individual responsible.

15.5 SERVICE EXAMPLE

One of the last tasks of NR and the network team is to document their work. Certainly, not every document generated is equally valuable, and Table 15.2 reflects this. This table describes the documentation set for the computing support

Table 15.2 Documentation set for computing support

Tree diagram – hardware, software, net support functions
Matrix – allocation of support to users
Operational definitions – maintenance by user type
Deployment chart – upgrade of department resources
Arrow diagram – connect new department to network
PDPC – developing an information server
Target/means matrix – area computing resource coordinator
Cascade diagram – server updating responsibility
Deployment matrix – deploy expansion review

expansion project. The affinity diagram is not listed, because it was superseded in terms of implementation by the tree diagram. Elements should be included in this list because they document the project in its final form, or because they are needed to document functions that continue past the end of the project.

15.6 SUMMARY OF KEY POINTS

- Match updating frequency to the criticality and time line of the control point's activity.
- Remember that reporting and updating should not necessitate face-to-face meetings.
- Avoid the dreaded 'meeting without a purpose'.
- Have a plan to deal with critical activities that become problems.

Afterword

Are we there yet?

(one of the authors at an early age, to the dismay of his parents)

Planning, especially in the sense of system and process improvement, is never quite finished, so we are never quite there. The project may be declared complete, but new processes have been developed and put into place. The alignment that was developed during the planning process needs to be standardized and institutionalized. Work continues, and lessons learned can continue to be applied, and spread into new areas.

The authors have tried to attack the planning process from a quality management point of view, but this could just as easily be viewed as a different insight into quality management. The two are not distinct, and if the trend to de-centralization and group work continues, they will be more intertwined. An example of this is in Appendix B, which treats the implementation of a TQM program as a project planning process.

This book began in the conviction that planning and quality are intimately related. It was motivated in part by people in companies which were trying to implement quality improvement programs, frequently with no clear idea of how to proceed. We have probably not answered their questions as clearly as we could, or as clearly as they might want. However, it was not our intent to present a 'cookbook' approach to quality or planning; a 'cookbook' approach would not work.

Rather than a cookbook, we present, in part, a book of recipes from which a meal can be constructed. Not every meal will contain every recipe; they are selected with care for the particular purpose. Just as important as the recipe is the context. The examples that tie these methods together provide a context that we hope helps understanding. Both the whole and the pieces are important. The pieces work, and when put together within an integrated plan, they achieve much.

APPENDIX A

Planning tool usage

A common difficulty in beginning use of TQM tools is connecting the tools with their use. Workers learn to use the tools, but are uncertain when or where to use them. Much of this knowledge comes from experience. It may be helpful, however, to begin with some perspective on how they are used. Table A.1 shows one arrangement of TQM tools as connections between inputs and outputs of a process.

Table A.1 Typical usage of TQM tools

Input	*Tool*	*Output*
Participation	Aha chart	Lessons learned
Jumble of ideas	Affinity	Structure
Related ideas	Tree	Hierarchy
Related ideas	Influence	Key logs
Actions	Flowchart	Order
Major activities	Top-down flowchart	Macro order
Outcome and causes	Cause and effect	Causal structure
Metrics	Pareto	Vital few
Related things	Matrix	Relationships
Key activities	PDPC	Contingencies
Key activities	PERT	Precedence
Goals, activities	Flag system	Tracking
Goals, activities	Targets/means	Alignment
Activities, people	Checksheet	Assignment
Components	The plan	Guiding document
Revisions	SDCA	Learning

APPENDIX B

TQM example

B.1 CONTEXT: EXISTING CORPORATE STRATEGY

The example presented here should be considered as the result of a series of activities engaged in by several people in a TQM implementation team and sub-teams. As such, it must be recognized that this is part of a larger process. We do not simply begin to implement TQM; TQM implementation exists within the larger context of a corporate strategy for continuous improvement.

The corporate strategy is very important, because corporate goals provide guidance for implementation. An attempt at TQM implementation that ignores or, worse, goes counter to overall corporate goals will be frustrating, counter-productive and highly likely to fail.

This example could have arisen from a small corporation or a major division of a large corporation. It assumes that other activities, such as strategic planning and expression of corporate commitment to TQM, are in place. It also does not address other kinds of activities that are frequently regarded as part of TQM, such as benchmarking and process improvement. As such, it is less than a full implementation of TQM. It focuses on one of the most difficult and important aspects of implementation, development and execution of a meaningful plan. This example is incomplete in that it does not display every result of every activity that goes into planning implementation. At each stage of work, there may (and frequently should) be additional activities and results, in particular in the stages when sub-teams begin deploying into their areas of work.

B.2 DEFINITION PHASE

The process begins with definition of a purpose within the context defined by the company. In some cases, where a well-articulated strategic plan exists, this may be a simple matter. In others, it requires that more effort be spent in translating the strategic plan and corporate context into terms and forms that will be easily understood. Both consensus and understanding are vital at all stages, and they begin here.

We begin with an affinity process. A good key question is 'What are the important factors for TQM implementation in a major department?' If we are dealing with a small business or corporation, this question would apply as well. We use this question to guide the affinity process, and bring out a consensus definition of the scope of the task.

An affinity diagram that could have resulted from this question is shown in Figure B.1. Note that there are both action-oriented items as well as concepts present. The affinity process develops as guided by the participants, and the results reflect this. In this case, we have one grouping that is explicitly conceptual. These can provide guidance in addition to the strategic plan and other guiding principles that may be in place. Many of the other groupings imply actions to be taken, such as providing training in TQM tools. These also provide some conceptual guidance, as in customizing training by need, and stressing the importance of executive training.

While the affinity diagram provides some organization, a tree diagram provides more, and is a more natural way to show system relationships. The tree diagram in Figure B.2 could have been developed by asking 'What additional activities fit within the affinity structure?', making it a guide for actions. An additional tree diagram could have been developed to display precedence of concepts and guidance, or

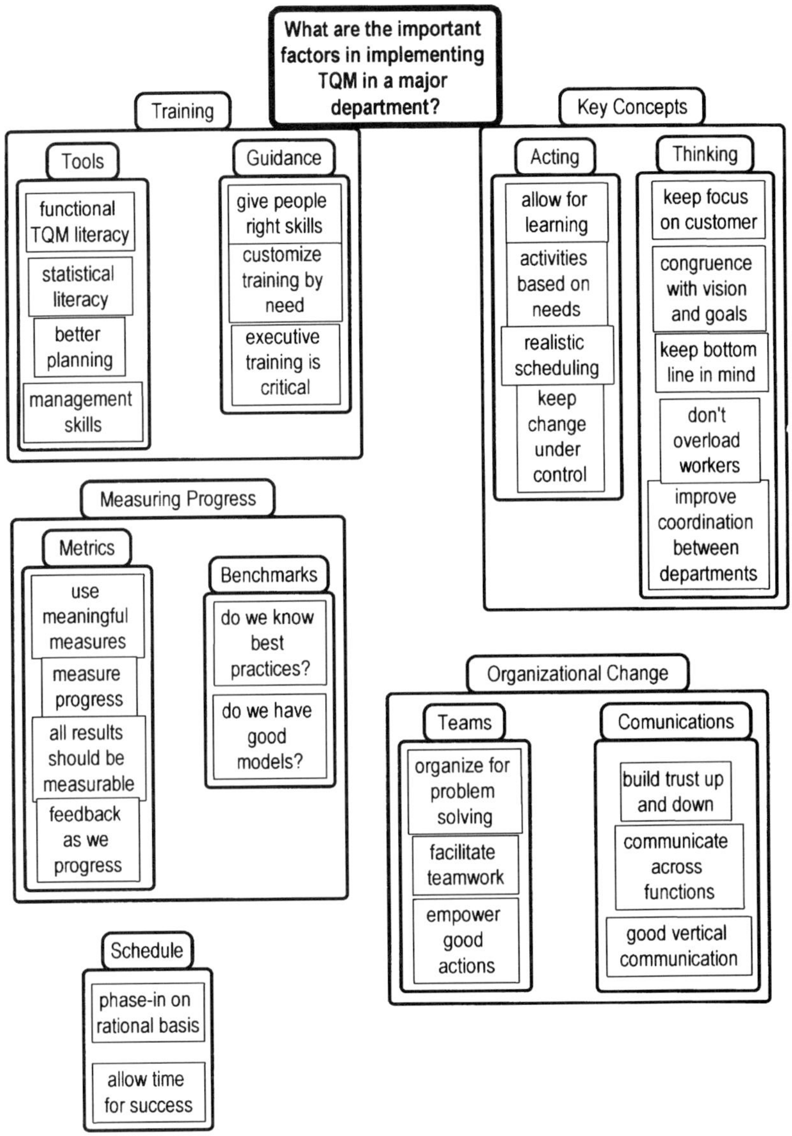

Figure B.1 Affinity diagram of overall TQM plan

they could be added as notes to this diagram. Some items from the affinity process have been subsumed together here, and others omitted because they don't fit this particular diagram. In

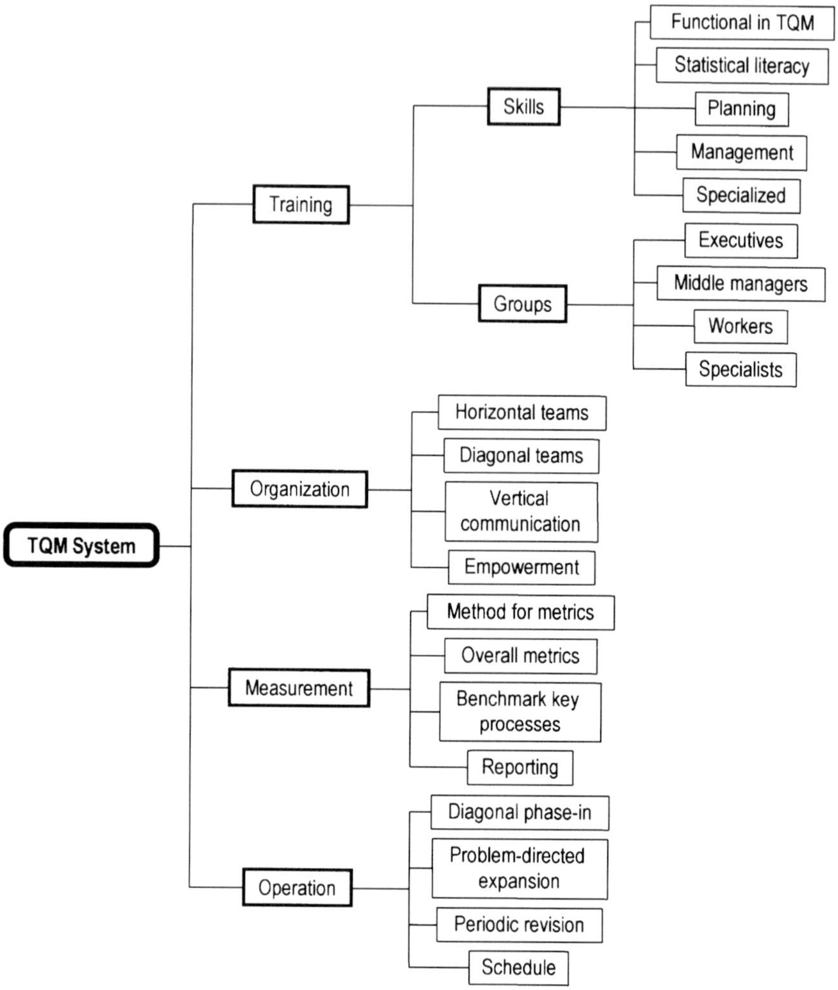

Figure B.2 TQM plan tree structure

addition, new items can be placed on the tree diagram as a result of thinking stimulated by the affinity diagram.

This illustrates one of the pitfalls in using examples from references or other companies in guiding TQM implementation. The examples show results, not the processes used to achieve them. Further, the examples arise from a particular set of circumstances and people, and probably would not be duplicated even if the same people were involved. Examples, including these, are best used as guidance in engaging in TQM processes. They provide information about the tools and processes, and facilitate implementation by answering some of the 'how-to' questions. Good planning and execution require both will and understanding. The will and commitment to proceed is easier to attain if the understanding of how to proceed is not a roadblock.

B.3 ANALYSIS PHASE

Having defined the context and overall structure, we proceed to analysis. At this point we have the rough outlines of the TQM system. We now begin to understand the characteristics of the system that is being developed. As we proceed through TQM processes that begin to add detail, we will address relationships, sequences of activities and priority, and begin to define the structure and characteristics of sub-systems.

The influence diagram is a particularly useful tool as we begin to deal with the many activities that are part of the TQM system. By beginning with influences, not necessarily precedence, we arrive at a structure that displays the interlocking of relationships between activities. Identifying potential 'key logs' or bottlenecks before they occur allows us to deal with them in a proactive rather than a reactive way.

This influence diagram is built up from action items taken from the affinity and tree diagrams. As with the tree diagram, the process of developing the influence diagram may result in items that are aggregated or re-defined. We may also discover new items that were overlooked or implied by previous processes.

The key activities in this diagram are identified from the number of arrows entering and leaving. Mid-manager training, for example, is important because it influences four other activities. Empowerment and diagonal phase-in, however, are important because they depend on (or are influenced by) several other activities. We don't simply use a rigid rule to locate bottlenecks. We use judgment in identifying them, and analyze them individually in determining how to cope with them.

The influence diagram begins to imply priority and precedence, but does not strictly define it. We take the influence structure and formalize it into a precedence structure in a top-down flowchart, as shown in Figure B.4. 'Top-down' in this context does not mean vertical flow. It means to develop the flow of major or top-level activities. Within each major activity we may identify sub-activities or implementation notes. We need to start at this point, rather than a detailed flowchart, because even for a simple process the details quickly become cumbersome to manage.

The complexity of detailed process flow is illustrated in the structure of one block in the top-down flowchart, development of a training plan. Part of this plan is shown in Figure B.5. Key words in the top-down flowchart correspond to segments of the training plan activity flow. This figure shows part of the detail, beginning with identifying training needs and key groups, and connecting into other charts for diagonal and horizontal teams.

As in previous processes, as we add detail we revise and modify what we know. The schedule group was apparently independent on the affinity diagram. It appeared under the operation heading in the tree diagram, not under the training heading. In the influence diagram, it stood between overall metrics and diagonal phase-in and reporting, and is mentioned under the initiate-training activity in the top-down flowchart. When we begin to plan details, however, we find a need to consider a training schedule in developing the training plan. If we had begun at the detail level, there could have been good cases presented on why schedule is important enough to develop early in the process. A likely result would have been time consumed in attempting to develop a schedule before enough information was available. By proceeding logically, from concepts through major activities to detail, discussion of scheduling will appear naturally in several places, at points in the overall process where schedule can be intelligently discussed.

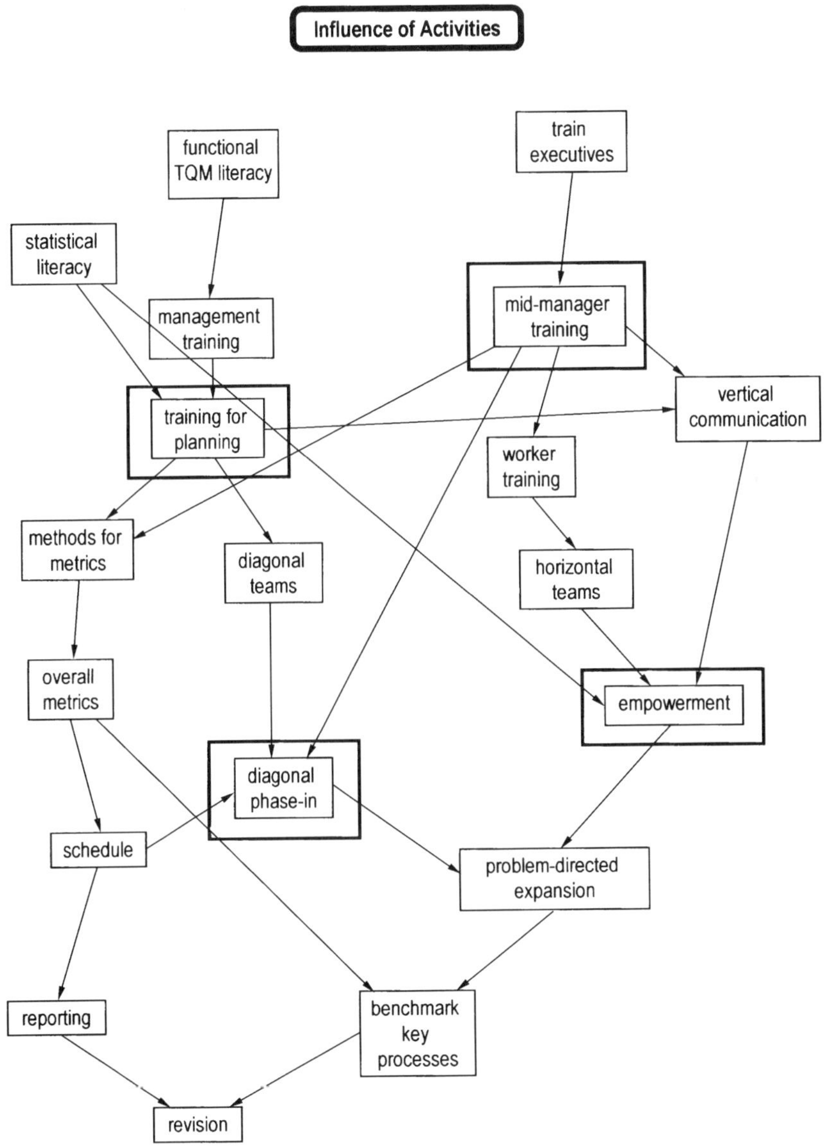

Figure B.3 TQM plan influence diagram

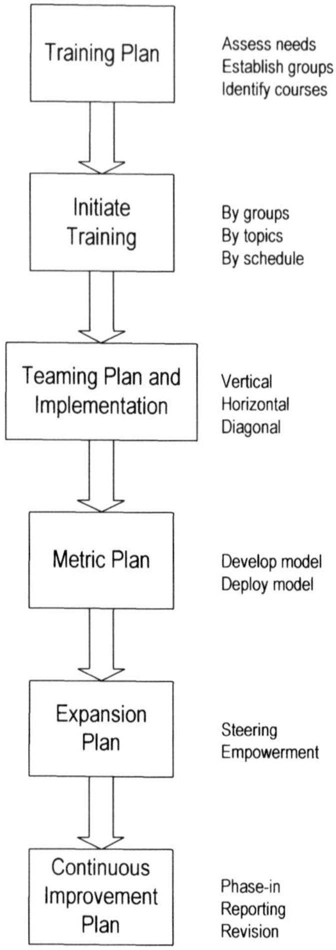

Figure B.4 TQM plan top-down flowchart

In a complete TQM process, more activity flowcharts would be developed at this point, either by the lead group or by sub-groups. The most effective way to proceed would be to provide the top-down flowchart and supporting documentation to sub-groups which are knowledgeable in implementation details, and likely to be responsible for implementation

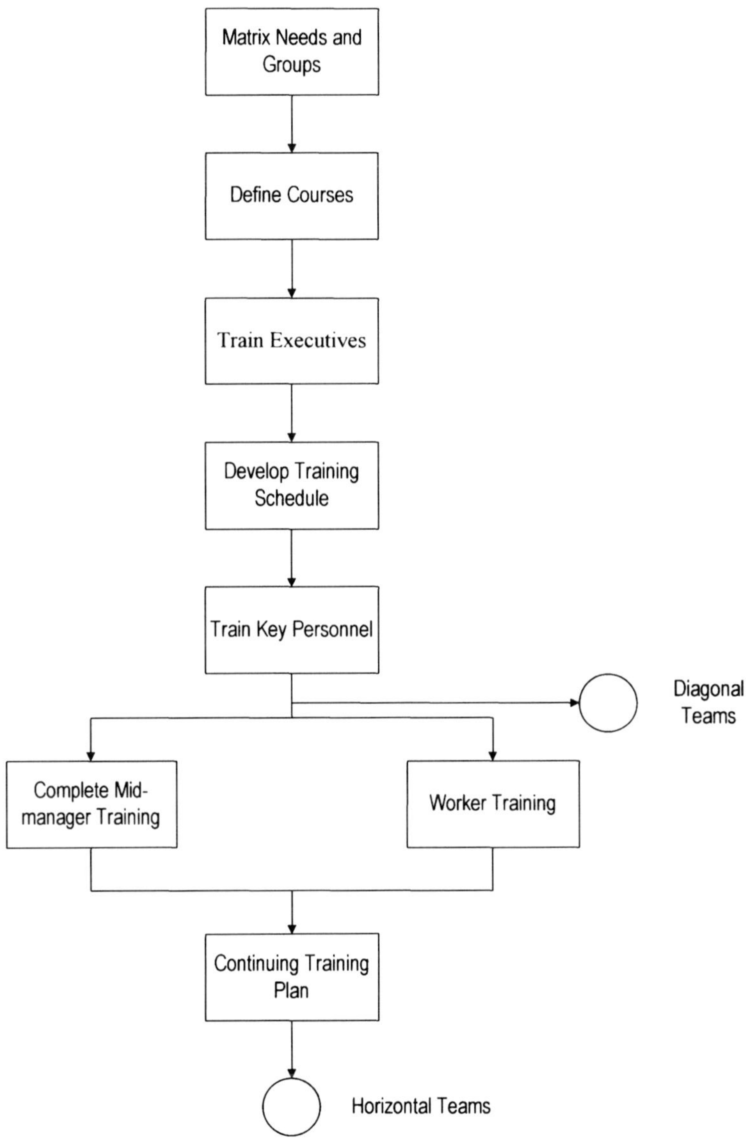

Figure B.5 Activity flow for training plan

and execution of pieces of the overall plan. The lead group then moves on into a new perspective on analysis.

As sub-groups begin work, the need for resources beyond the core team arises. Work begins to expand in sub-groups, which will need some guidance. Sub-groups may also develop conflicting goals, and contention for resources. It is important to be able to provide guidance and leadership to the sub-groups. For this to happen, the lead group must consider all major activity areas and overall goals. This leads to a shift from concepts and overall flow to detail and prioritization.

One very useful tool at this point is the Pareto chart. For example, the group may ask 'What major areas within the training plan do we address first?' This can be answered in a number of ways, depending on the objective. One possible objective is to develop self-sufficient teams within departments. Another is to get all personnel trained. The Pareto chart in Figure B.6 could help prioritize departments if we want to merge these objectives by developing self-sufficient teams, while training as many people as possible.

There are many other possible approaches. If we have good information on cost of poor quality, we could develop a Pareto chart on these costs by area. We could attempt to maximize impact on customers by using number of customers affected, or frequency of customer inquiries. It is desirable at this point in the analysis phase to develop several Pareto charts, each displaying a different dimension or set of objectives. Overall strategic goals or guiding principles will help decide among the objectives, and help guide the early direction or emphasis in implementation.

Continuing to add detail to the plan, we now consider a different aspect of the plan. The Pareto charts provide prioritization of activity. Implementation of an activity requires that we understand the causal structure within it. It is a truism that we cannot do everything at once. Therefore, what we do must be directed well, and one aspect of this is connection of causes with desired results. The cause and effect diagram is useful here.

Figure B.7 shows a causal structure for a very common

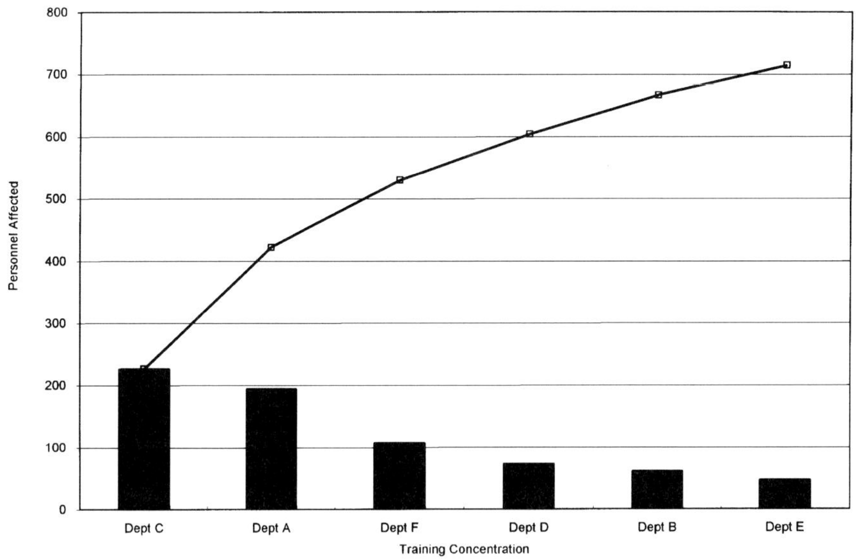

Figure B.6 Pareto analysis of department size

problem in TQM implementation – the ability to function within teams. If the organization is not experienced in teamwork, and many are not, the ability to team is a critical skill that must be taught. With 'ability to team' as a desired effect, we enumerate potentially related causes in logical groups. The resulting diagram can provide guidance in developing a training syllabus for teaming.

The cause and effect diagram is most often used in process improvement. The classic use is with an effect in terms of poor quality, and causes taken from the physical process. Another use has a positive effect, with enabling or facilitating factors as causes. Its use here illustrates its flexibility. In all cases, the diagram is a starting point, with other tools and methods used to further develop decisions regarding causes. In the example here, knowledge of the existing skills of the workforce would lead to a prioritization of emphasis (perhaps using a Pareto chart) in developing the training syllabus.

Another dimension of added detail in analysis is simply

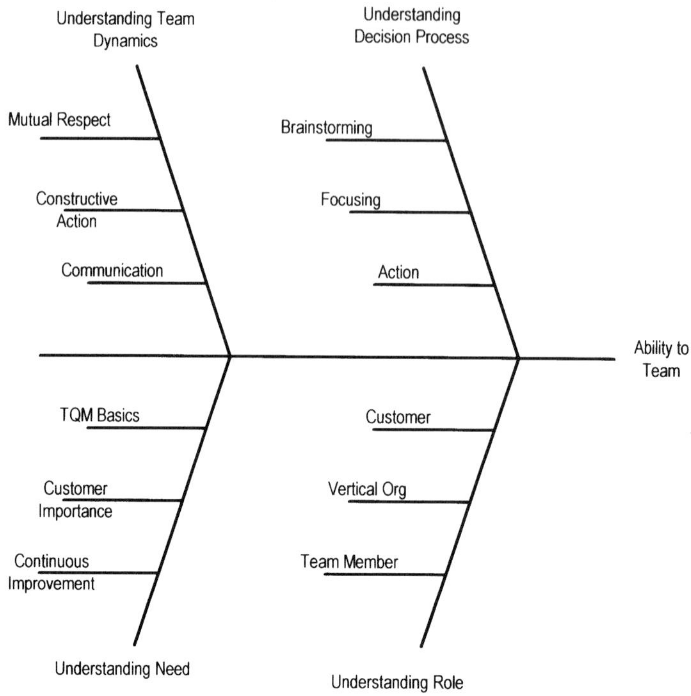

Figure B.7 Teaming ability causal structure

representation of what we know. A highly useful and versatile tool for data representation is the matrix diagram. As the core group and sub-groups work to add detail to the analysis, it is important to document knowledge as it is developed, and present it as simply as possible. Figure B.8 illustrates this.

As sub-groups have worked to add detail in training as guided by the top-level plan, specific courses are being developed. The causal structure for training leads to identification of critical skills. The matrix shows one possible set of relationships among training populations, critical skills, and potential courses. The filled circles show strong relationships, and the empty circles show moderate relationships. Weak relationships, if present, would be represented by triangles. Figure B.8 shows us how specific courses (TQM Basics, problem-solving

PROBLEMS/ ACTIVITIES	TQM Basics	7QC Tools	Prob Solving	PDCA SDCA
Work Group Training	⊙		⊙	
General Training	O	⊙		⊙
Individual Training		O		O
Team Dynamics	⊙			
Team Decisions	⊙		O	
Need for Teams	⊙			
Role in Teams	⊙			
Statistical Thinking		⊙		O
Customer Orientation	⊙			
Methodical Approach			⊙	O
Self Management			⊙	⊙

Figure B.8 Relationships among courses, groups and topics

etc.) are related to both training populations and critical skills. Another useful matrix would show relationships between training populations and critical skills.

B.4 SPECIFICATION PHASE

This phase is another addition of detail to the plan, but in a different sense. Deployment into sub-groups began in the analysis phase, and the added detail was primarily description of components of the system, to allow them to be developed. In this phase, the added detail is the specification of the components. We describe them in more detail, ideally in sufficient detail to allow them to begin operation.

Operational definitions should provide sufficient detail to avoid confusion in implementation. Several formats are possible; the one used here is a matrix format, as shown in Figure B.9. The operational definition of each key activity is constructed using the four Ws and one H, followed by 'Ask why five times'. In this example to define the details of worker training, the 4W1H columns identify the logistical details of worker training. The 5Why column is a reminder to engage in the exercise. An attachment of the results of this exercise would be valuable in ensuring that the training achieves its goals.

Operational Definition Worker Training						
	Who	What	When	Where	How	5Why
TQM Basics	All	Std TQM	Monthly	Cent Tng	Lec/Lab	*
7QC Tools	All	Std QC	Monthly	Cent Tng	Lec/Lab	*
Team Problem Solving	All	Custom	Qtrly	Local Tng	Fac Grp	*
PDCA/SDCA	All	Custom	Qtrly	Local Tng	Fac Grp	*
Specialty	As Req'd	Define	As Needed	Define	Define	*

Figure B.9 Operational definition for worker training

An especially important specification is the description of processes involving more than one person or group. The deployment chart was developed for this function, and serves to document the process well. The chart in Figure B.10 is for the process of training new employees in TQM. While the majority of training, especially early in the process, will be dedicated to existing employees, we must be sure to institutionalize the processes to ensure continuation. If we don't develop this process quickly, new employees will be left out, and depending on the personnel turnover rate, a severe training demand will develop later.

This process assumes an initial interview between the new employee and supervisor, followed by an orientation provided by the TQM Training area. Training in specific skills is determined by the supervisor and the Skills Training area. Following the training, a critique is done to decide whether more training is needed. If training is complete, a mentor is assigned. Periodic capability reviews with the employee and supervisor are used to determine need for additional training. The deployment chart documents this well, and is easily understandable by both supervisors and workers.

This kind of specification documentation would be developed for all critical processes in this phase. The methods are easy to learn, and most of this work is done by sub-groups having responsibility for the functions.

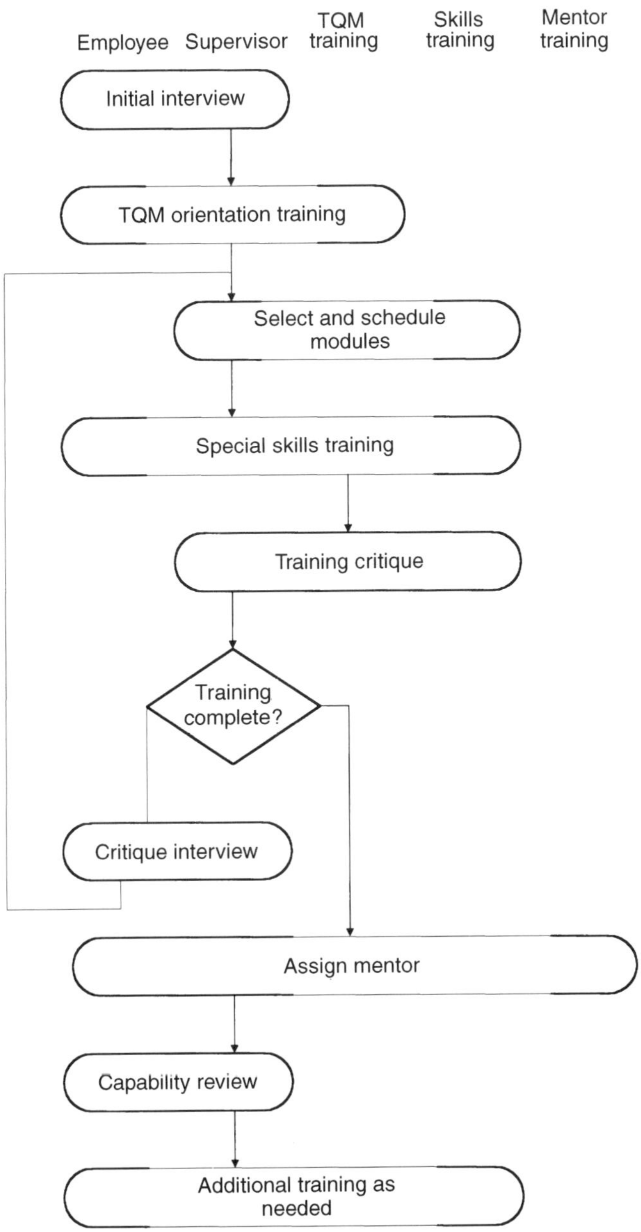

Figure B.10 Deployment chart for new-employee training

B.5 DEVELOPMENT PHASE

The analysis phase identified components of the system, which were then described in detail. At this point we have key activities defined and specified. This stage is a potential source of problems, because the effort is broadening across functions and involving more people. It is important that deployment does not turn into chaos. To avoid this, we formalize the structure of the previously defined components. If we can understand their relationships, we can guide their implementation.

A familiar tool for describing a routine and repeatable process is the arrow network (Figure B.11). With this tool we describe the precedence, and optionally the resource requirements, in a set of activities that make up a complete process. This differs from the deployment chart in that the emphasis here is precedence, and each activity is typically the responsibility of a single individual or group. There may be several groups involved in the overall process, however, and the network serves to identify information sources and customers, as well as precedence. If resources are also called out on the network, a group can balance several resources required in several processes.

This arrow network describes the process of developing a training plan. Since a different training plan will be needed for each major group of employees, it is worth the effort to document the process. In some ways this network is similar to the activity flowchart in Figure B.5. That flowchart, however, was used at a higher level, and each block in it could correspond to its own arrow network. This use of a hierarchy of flowcharts and arrow networks is preferred to a single large diagram, which can easily become overwhelming and uninformative.

Non-routine activities, or processes that have not been done before, need a different kind of chart. If the process is new, there may be unforeseen difficulties in it. Certainly, it has not become routine, and we should not expect everyone involved in it to share a common understanding. More importantly, we need to ensure, to the extent possible, that the unforeseen difficulties do

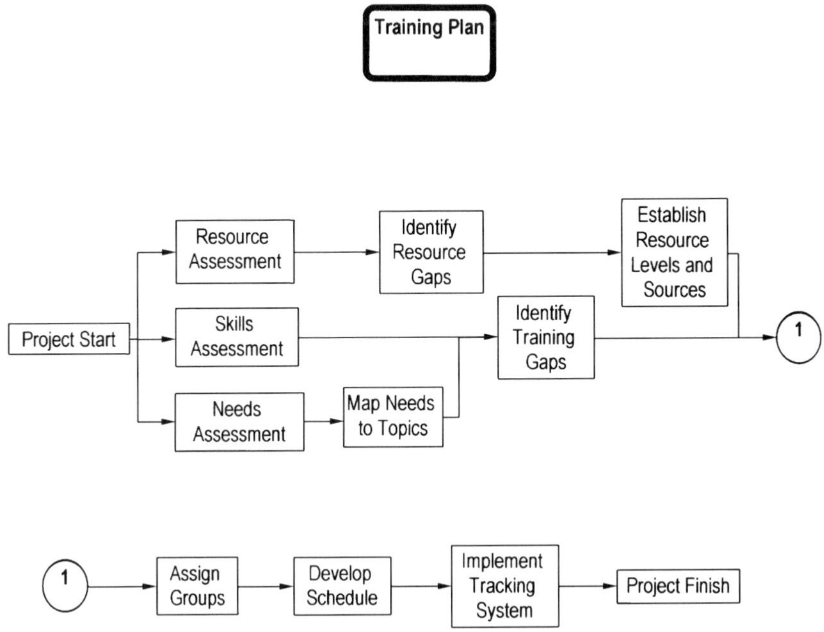

Figure B.11 Activity sequence in training

not stop the process. This is guarded against by use of the process decision program chart (Figure B.12).

This chart is designed to help organize things that can go wrong and contingencies against them. The top level in the chart is a set of activities. These have come from previous processes in TQM planning, related to the formation of teams. The second level, developed through brainstorming, is a set of potential obstacles or things that could go wrong. The rounded boxes are possible contingencies against the obstacles. Once they are identified, each is evaluated for desirability. The desirable contingencies are marked with an 'O' and the undesirable ones with an 'X'. As in many other TQM tools, the chart is not the process, it is documentation of the process in a useful form. The process in this case has taken place within the group responsible for the teaming plan, and has probably used other tools, such as brainstorming and affinity diagrams, to develop the data in the

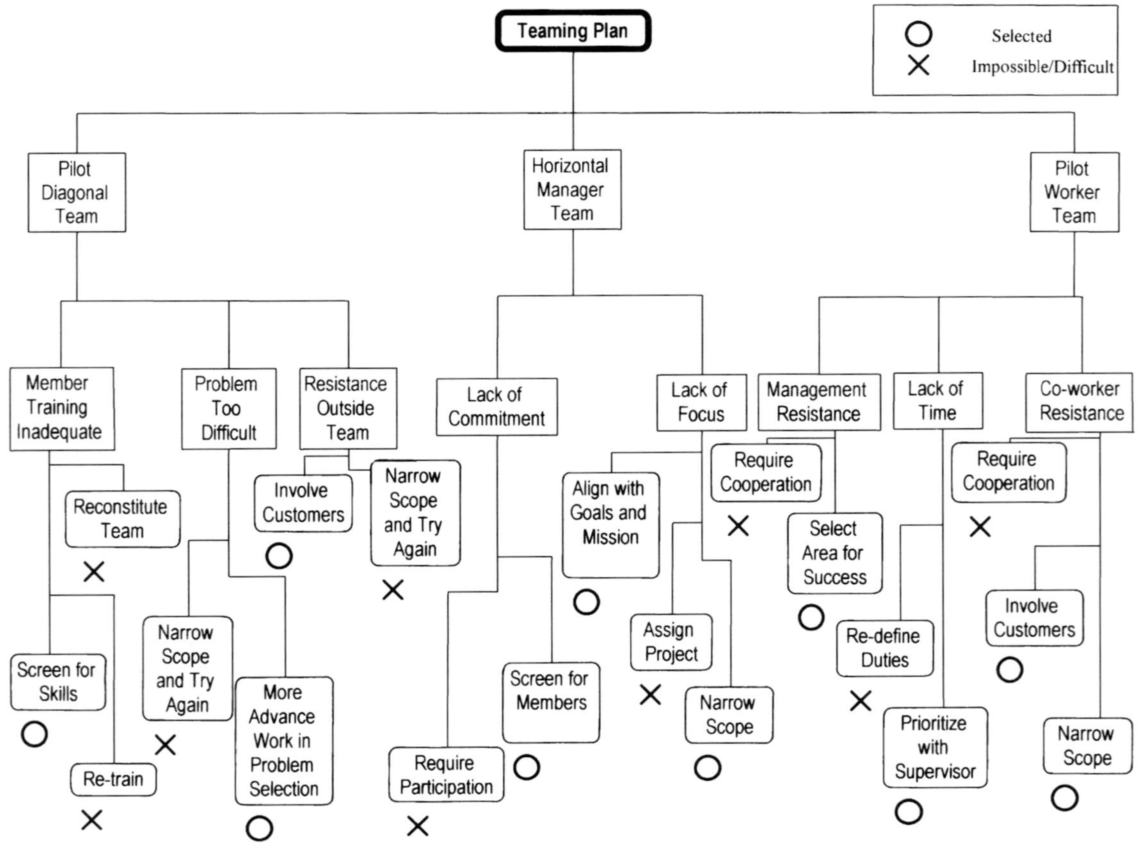

Figure B.12 Contingency structure in teaming plan

chart. The chart then provides guidance in making decisions about actions to be taken.

The development phase will result in development of several arrow networks and PDPCs. It is very important, as deployment of the TQM plan proceeds into new areas, that the processes be well planned. The use of these tools involves teamwork among the people who will be key players in TQM implementation. Their involvement in planning implementation by documenting critical processes will be crucial to the overall success of the plan.

B.6 IMPLEMENTATION PHASE

The implementation phase assumes that key processes and functions have been defined and specified. It is neither necessary nor likely that all processes and functions in the organization have been dealt with. The key processes, however, are being implemented. This is the second critical phase of the TQM plan. It is very easy at this stage to believe that we are almost finished. If this happens, the entire effort loses focus. The key to successful implementation is to ensure that activities continue to be aligned with strategic goals, and that progress continues to be measured in terms that are also consistent with strategic goals. In other words, at this stage we link top-level objectives with how they are satisfied at lower levels.

A simple example of strategic alignment is the targets/means matrix, shown in Figure B.13 for a middle-level manager's role in the system. The targets are taken from one level up. In this case, the manager's progress is being evaluated by his personal training, his employee training, and the extent of late completion of

Mid-Manager Progress		Targets			Check Points		
		Personal Training	Employee Training	Project Lateness	Number	Item	Data
Means	Personal Training	⊙		△	C1001	Cont Ed Hrs	Bar Chart
	Subordinate Training		⊙	○	C1002	Cont Ed Hrs	Bar Chart
	Quarterly Review		△	○	C1003	Pers Rev'd	Check Sheet
Control Points	Number	R1001	R1002	R1003			
	Item	Cont Ed Hrs	Cont Ed Hr	Mthly Avg			
	Data	Bar Chart	Bar Chart	Line Graph			

Figure B.13 Targets and means for a middle manager

projects. These would be taken from the goals of the manager one level higher in the organization, and all would be components of strategic objectives. Their control points are identified, as well as how they are to be tracked.

The means to success in achieving these targets are selected by the manager, in this case personal and subordinate training and quarterly project reviews. The checkpoints monitor the input side of the manager process. Relationship symbols in the matrix show how the inputs affect the outputs. It is interesting to see the importance of training in project lateness, recognizing that the manager cannot force success, he can only facilitate it by ensuring that his workers have the correct tools.

One level up could find a manager using a flag system to track TQM training, among other things. The simple flag system in Figure B.14 gathers data on employees trained by department, and aggregates it into the number trained in the division. This manager, who probably supervises several managers at the level in the targets/means example, might use several flag systems to cover the key measurable activities in his area of responsibility.

These charts go by a variety of names. They are found in Hoshin planning, sometimes called policy deployment, and are related to the original form of management by objectives. What all of these have in common is a translation of strategic objectives from level to level. At each level of the organization, strategic objectives are translated into meaningful terms for that level. Components of the strategic plan that are implemented at that level are explicitly identified. Finally, a measurement system is put into place that allows the people involved to track their own work by monitoring their inputs at checkpoints, monitoring their results at control points, and understanding their roles in the overall organization.

The cascade diagram in Figure B.15 illustrates this for two levels. Targets at the upper level are translated into targets at the next level down, and control points are identified. The means to achieve those targets at the upper level are also translated down, and implemented at the lower level in terms appropriate to that level. This simple deployment and monitoring system, designed as it is for use by the people involved, is necessary to

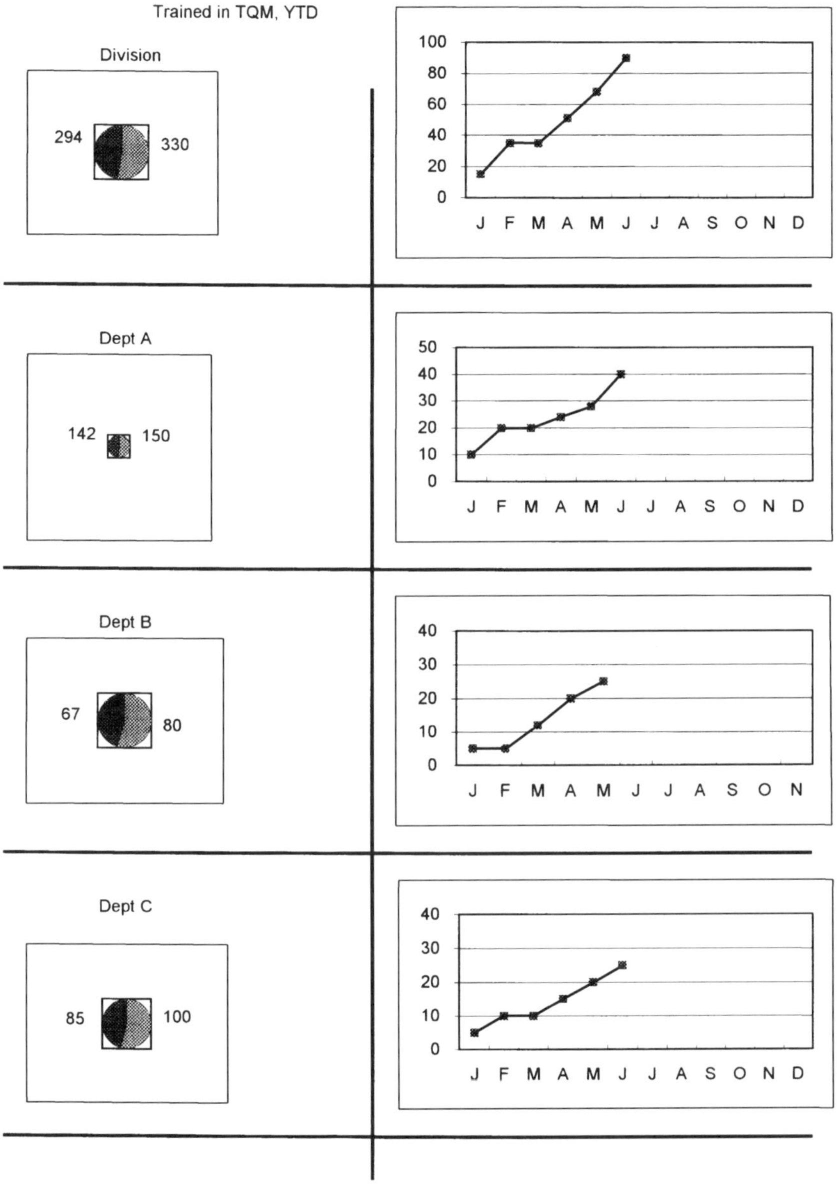

Figure B.14 Tracking departmental training by flags

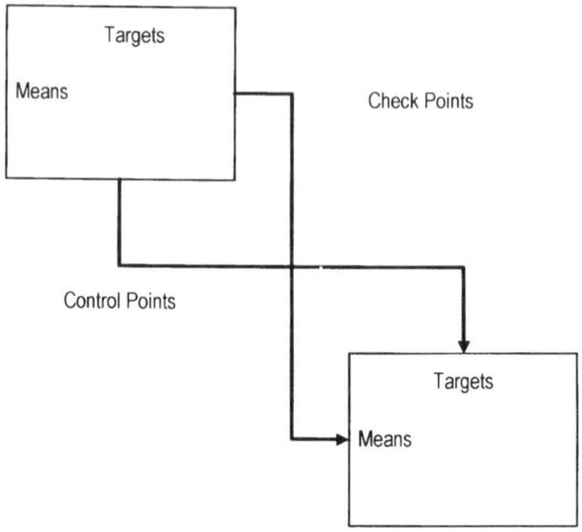

Figure B.15 Cascading the implementation

ensure that the overall set of activities results in satisfying top-level goals.

B.7 REVIEW PHASE

The review of TQM implementation actually starts before implementation is complete. In this phase we associate the planning level with the implementation level. The implementation needs to be guided by monitoring and planning for use of resources, for example, using the matrix in Figure B.16. As the implementation expands, simple things become more difficult due to the scope of the activities, the number of people involved, and the resources they require. We begin our review by planning to accommodate the expansion.

	Instructor	Facilitator	Tng Rm "A"	Tng Rm "B"	Coordination	Scheduling
TQM	⊙			⊙		
7QC Tools	⊙		⊙			
Problem Solving	⊙	O		⊙		
PDCA/SDCA	⊙	O	⊙			
Mentors		O		△	⊙	
Specialty	△			△		△

Figure B.16 Tracking resources and needs with a matrix

Table B.1 Key TQM planning tools

Affinity process
Tree diagram
Influence diagram
Flowchart
Pareto chart
Cause and effect diagram
Matrix diagram

Table B.2 Key TQM plan documentation components

Deployment chart
Arrow diagram
Process Decision Program Chart
Targets/means matrix
Operational definitions

It is also very important to ensure that all involved are working from the same set of plans. Now that the planning process is complete, it needs to be documented. This not only provides a template for future activities, it facilitates sharing of information among the functions involved. The set of tools in Table B.1 clearly document the planning process. The affinity, tree, and influence diagrams apply to the overall process. As the tools become more detailed and specific to areas, it becomes useful to develop sets of flowcharts, Pareto charts, cause and effect diagrams, and matrices to document specific functions.

The middle-level planning process, where implementation takes place primarily within functional groups, is documented by the more detailed tools in Table B.2. This is the operational level, where it is vital that all involved share a common understanding of processes. The deployment chart, arrow diagram, and operational definitions document the processes as they are to operate. The other tools provide useful supporting information for implementation.

Table B.3 Key TQM plan implementation tools

Flag system
Cascade diagram
Targets/means matrix
Deployment chart

The methods used in the implementation phase for alignment, found in Table B.3, provide the information for implementation review. The targets/means matrix, in particular, identifies measures of performance within each department that can be monitored through flag systems to provide feedback on the overall implementation. These provide documents that will be used in tracking further implementation, as well as evaluating success in implementation.

Having defined the mechanisms for review, there is still some implementation work to be done. Learning takes place at each step of the way, and we need to take advantage of it. A flowchart of the overall process is shown in Figure B.17. It is highly useful to loop back at least once from reconciling information through deployment and implementation. First, this provides an opportunity to take advantage of learning in a second directed phase of implementation into new areas. This loop also allows us to revise what we have done to improve it. While the steps in this implementation process are depicted as sequential, they are not completely independent. Just as designs of products are improved by iteration after review, the design of a TQM system can also be improved by guided iteration.

This loop back through can also identify processes in need of attention, or needing more detail for implementation. Continuous improvement applies to TQM systems themselves, as well as to the processes that are the explicit targets. It is unreasonable to expect that one pass through each function will result in fixing it forever.

Having made at least one more pass through the deployment and implementation loop, it is now time to begin tracking

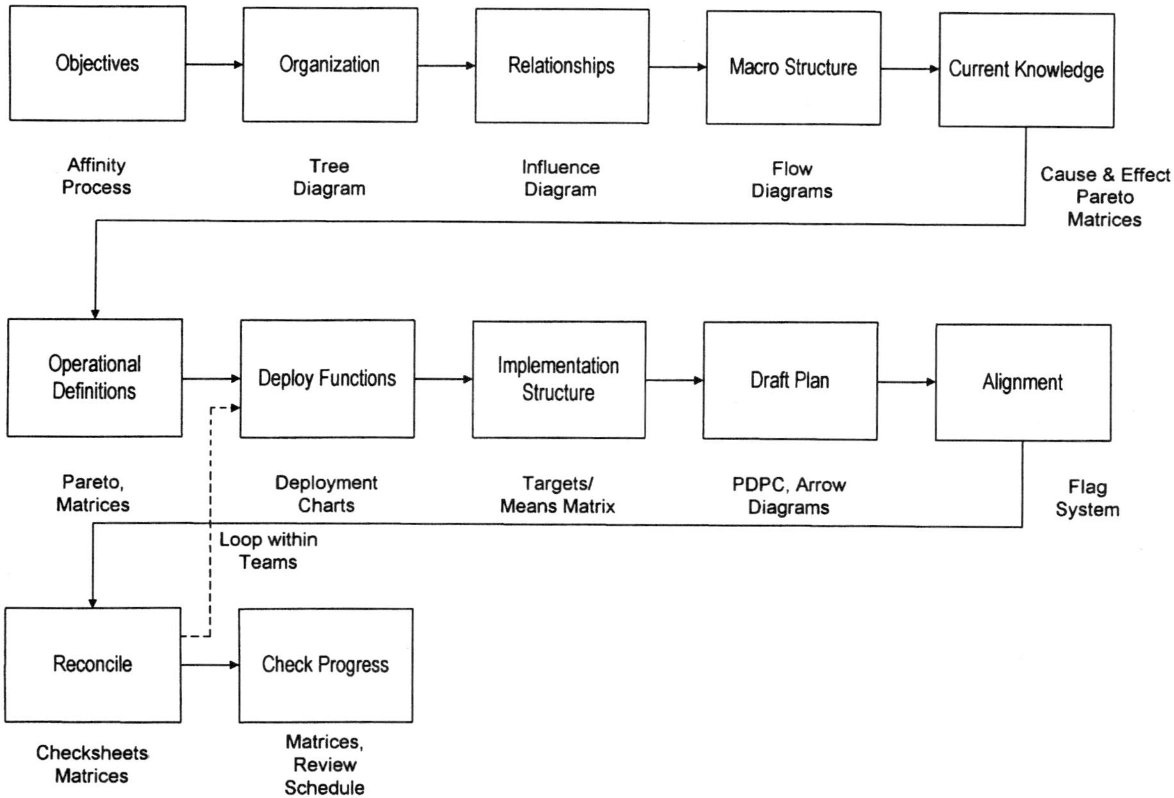

Figure B.17 Looping through the planning process

progress formally. The review and tracking process needs to be designed just as any other process. A tracking schedule must be developed and reporting requirements distributed. Reporting responsibility must also be identified, and it is important here that the people busily implementing TQM not be unduly burdened with paperwork.

Finally, having made one complete pass, there is still work to be done. At this point, the implementation process should be well understood. It is time now to deploy into smaller teams, and begin new loops from the analysis phase in new areas, continuing the expansion of the implementation.

Selected bibliography

Akao, Yoji (1991) *Hoshin Kanri*, Productivity Press, Cambridge, MA.: the basic reference on policy deployment and tools such as flag systems and target/means matrices.

Brassard, Michael (1986) *The Memory Jogger +*, GOAL/QPC, Methuen, MA: an excellent working reference on the Seven Management and Planning Tools.

Brassard, Michael and Diane Ritter (1994) *The Memory Jogger II*, GOAL/QPC, Methuen, MA: a very convenient (pocket-sized) reference on the Seven Quality Control Tools, the Seven Management and Planning Tools and other handy methods.

Deming, W. Edwards (1995) *The New Economics*, MIT Center for Advanced Engineering Studies, Cambridge, MA: Dr Deming's most definitive work on the System of Profound Knowledge.

Galgano, Albert (1994) *Companywide Quality Management*, Productivity Press, Cambridge, MA: a broad reference on quality management.

Imai, Masaaki (1986) *Kaizen, the Key to Japan's Competitive Success*, Random House, New York: the seminal reference on continuous improvement, by the originator of Kaizen.

Ishikawa, Kaoru (1990) *Introduction to Quality Control*, 3A Corporation, Tokyo (English translation): a significant expansion of Ishikawa's Guide to Quality Control which includes material on planning and the quality function.

Juran, J. M. (1992) *Juran on Quality by Design*, The Free Press, New York: an excellent reference on planning and quality planning, more than the title might indicate.

Mizuno, Shigeru (1988) *Management for Quality Improvement – The 7 New QC Tools*, Productivity Press, Cambridge, MA: the basic reference on the Seven Management and Planning Tools.

Moder, Joseph J., Cecil R. Phillips and Edward W. Davis (1983) *Project Management with CPM, PERT and Precedence Diagramming*, Van Nostrand Reinhold, New York: a good basic reference on traditional project planning.

Shiba, Shoji, Alan Graham and David Walden (1993) *A New American TQM*, Productivity Press, Cambridge, MA: TQM as a total system, including planning and policy deployment.

Tribus, Myron (1989) *Deployment Flowcharting*, Quality and Productivity Inc., Los Angeles: a short monograph with the basics of deployment flowcharting and some examples.

Index